Natural Fiber-Reinforced Hybrid Composites

Natural Fiber-Reinforced Hybrid Composites

Special Issue Editor

Vincenzo Fiore

MDPI • Basel • Beijing • Wuhan • Barcelona • Belgrade • Manchester • Tokyo • Cluj • Tianjin

Special Issue Editor
Vincenzo Fiore
University of Palermo
Italy

Editorial Office
MDPI
St. Alban-Anlage 66
4052 Basel, Switzerland

This is a reprint of articles from the Special Issue published online in the open access journal *Fibers* (ISSN 2079-6439) (available at: https://www.mdpi.com/journal/fibers/special_issues/natural_fiber_reinforced_hybrid_composites).

For citation purposes, cite each article independently as indicated on the article page online and as indicated below:

LastName, A.A.; LastName, B.B.; LastName, C.C. Article Title. *Journal Name* **Year**, *Article Number*, Page Range.

ISBN 978-3-03928-154-1 (Pbk)
ISBN 978-3-03928-155-8 (PDF)

© 2020 by the authors. Articles in this book are Open Access and distributed under the Creative Commons Attribution (CC BY) license, which allows users to download, copy and build upon published articles, as long as the author and publisher are properly credited, which ensures maximum dissemination and a wider impact of our publications.
The book as a whole is distributed by MDPI under the terms and conditions of the Creative Commons license CC BY-NC-ND.

Contents

About the Special Issue Editor vii

Preface to "Natural Fiber-Reinforced Hybrid Composites" ix

Matthew Chapman and Hom Nath Dhakal
Effects of Hybridisation on the Low Velocity Falling Weight Impact and Flexural Properties of Flax-Carbon/Epoxy Hybrid Composites
Reprinted from: *Fibers* **2019**, *7*, 95, doi:10.3390/fib7110095 1

Le Quan Ngoc Tran, Carlos Fuentes, Ignace Verpoest and Aart Willem Van Vuure
Tensile Behavior of Unidirectional Bamboo/Coir Fiber Hybrid Composites
Reprinted from: *Fibers* **2019**, *7*, 62, doi:10.3390/fib7070062 17

Pietro Russo, Giorgio Simeoli, Libera Vitiello and Giovanni Filippone
Bio-Polyamide 11 Hybrid Composites Reinforced with Basalt/Flax Interwoven Fibers: A Tough Green Composite for Semi-Structural Applications
Reprinted from: *Fibers* **2019**, *7*, 41, doi:10.3390/fib7050041 27

Fabrizio Sarasini, Jacopo Tirillò, Luca Ferrante, Claudia Sergi, Pietro Russo, Giorgio Simeoli, Francesca Cimino, Maria Rosaria Ricciardi and Vincenza Antonucci
Quasi-Static and Low-Velocity Impact Behavior of Intraply Hybrid Flax/Basalt Composites
Reprinted from: *Fibers* **2019**, *7*, 26, doi:10.3390/fib7030026 37

Mohammad Bellal Hoque, Solaiman, A.B.M. Hafizul Alam, Hasan Mahmud and Asiqun Nobi
Mechanical, Degradation and Water Uptake Properties of Fabric Reinforced Polypropylene Based Composites: Effect of Alkali on Composites
Reprinted from: *Fibers* **2018**, *6*, 94, doi:10.3390/fib6040094 53

Muhammad Ahsan Ashraf, Mohammed Zwawi, Muhammad Taqi Mehran, Ramesh Kanthasamy and Ali Bahadar
Jute Based Bio and Hybrid Composites and Their Applications
Reprinted from: *Fibers* **2019**, *7*, 77, doi:10.3390/fib7090077 63

About the Special Issue Editor

Vincenzo Fiore has been Assistant Professor in technology and material science at the University of Palermo since August 2017. He graduated with honors in material engineering from the University of Messina in July 2004 and wrote his PhD dissertation on "Economic analysis, technological innovation and management for territorial development policies" for the University of Palermo in April 2008.

His research interest is focused on fiber-reinforced composite materials, with the main following topics:
- manufacturing and testing of composite materials
- extraction and characterization of new lignocellulosic fibers to be used as reinforcement of polymeric matrices
- manufacturing and testing of adhesive, mechanical, and mixed joints between similar and dissimilar materials
- manufacturing and testing of new eco-friendly materials with enhanced insulating properties
- evaluation of aging resistance of composite structures in hostile environments
- analysis of viscoelastic behavior of metal, glass, composite structures and natural materials

He is author or co-author of more than 50 publications in peer-reviewed journals, 2 patents, 3 book chapters and more than 40 conference presentations, seminars, and invited lectures. He has supervised or co-supervised more than 40 Masters' theses and has more than 5 years of teaching experience.

Preface to "Natural Fiber-Reinforced Hybrid Composites"

Due to their specific properties, low price, health advantages, renewability, and recyclability, natural fibers have received growing attention over the last few decades as an alternative to synthetic fibers used in the reinforcement of polymeric composites.

Nevertheless, natural fibers are hydrophilic, thus showing high susceptibility to moisture absorption and low resistance to humid and wet environmental conditions. Moreover, they show quite low and variable mechanical properties as well as weak adhesion with polymeric matrices. For these reasons, even if natural fiber composites are nowadays widely used in several industrial applications, including automotive, marine and infrastructure, their applications are limited to non-structural or semi-structural interior components.

In such a context, the production of polymeric composites reinforced with natural fibers together with synthetic counterparts can represent a valid applied compromise. This approach has been widely exploited in literature, and the resulting composites have shown a suitable balance of mechanical properties, thermal stability, aging tolerance against humid or aggressive environments, cost and environment care.

This book is comprised of five peer-reviewed original research articles and a review on jute-based hybrid composites. Topics include the investigation of quasi-static and low-velocity impact behavior of flax-carbon and intraply flax-basalt hybrid composites. In addition, the tensile behavior of unidirectional bamboo-coir fiber composites and the degradation and water uptake properties of polypropylene-based composites reinforced with pineapple-jute-cotton hybrid fabric were analyzed.

From these articles, it may be inferred that, based on their wide range of performance design, hybrid composites could emerge as a new alternative to engineering materials in several applications, which can optimize the use of synthetic laminates.

Hence, this volume could be useful for students as well as for designers and engineers who would like to develop a deeper understanding on the use of natural fibers with synthetic ones as reinforcement of composite structures.

Vincenzo Fiore
Special Issue Editor

Article

Effects of Hybridisation on the Low Velocity Falling Weight Impact and Flexural Properties of Flax-Carbon/Epoxy Hybrid Composites

Matthew Chapman and Hom Nath Dhakal *

School of Mechanical and Design Engineering, University of Portsmouth, Portsmouth PO1 3DJ, UK; matthew.chapman@myport.ac.uk
* Correspondence: hom.dhakal@port.ac.uk; Tel.: +44-23-9284-2582; Fax: +44-23-9284-2351

Received: 26 June 2019; Accepted: 14 October 2019; Published: 24 October 2019

Abstract: The trend of research and adoption of natural plant-based fibre reinforced composites is increasing, with traditional synthetic fibres such as carbon and glass experiencing restrictions placed on their manufacture and use by legislative bodies due to their environmental impact through the entire product life cycle. Finding suitable alternatives to lightweight and high-performance synthetic composites will be of benefit to the automotive, marine and aerospace industries. This paper investigates the low-velocity impact (LVI) and flexural properties and damage characteristics of flax-carbon/epoxy hybrid composites to be used in structural lightweight applications. LVI, for example, is analogous to several real-life situations, such as damage during manufacture, feasibly due to human error such as the dropping of tools and mishandling of the finished product, debris strikes of aircraft flight, or even the collision of a vessel with another. Carbon fibre has been hybridised with flax fibres to achieve enhanced impact and flexural performance. The failure mechanisms of woven flax and flax-carbon epoxy hybrid composites have been further analysed using Scanning Electron Microscopy (SEM). It was observed from the experimental results that carbon fibre hybridisation has a significant effect on the impact and flexural properties and their damage modes. The results obtained from this study exhibited that the flexural strength and modulus of plain flax/epoxy composite increase significantly from 95.66 MPa to 425.87 MPa and 4.78 GPa to 17.90 GPa, respectively, with carbon fibre hybridisation. This significant improvement in flexural properties would provide designers with important information to make informed decisions during material selection for lightweight structural applications.

Keywords: flax fibres; low-velocity impact; hybrid composites; mechanical properties; damage mechanisms

1. Introduction

The rise in global warming and the increased public awareness of the impact of pollution arising from the use of non-renewable sources is driving governments and business sectors to tackle climate change. There are many initiatives undertaken to stabilise and reduce the impact of greenhouse gasses (GHGs) on the natural world. The European Union (EU) for example has set penalties in the form of registration premiums [1] for all new vehicles registered, which exceed emission targets.

Natural fibres have a lower density and problem-free disposal, leading to them being a strong emerging alternative to synthetic fibres [2].

Composites Evolution [3] have produced a car door using a carbon/flax hybrid system. The company suggests that the mechanical properties of the carbon fibre are not significantly lost in a system where the inner layers of the composite structure are replaced with flax fibre. On the contrary, the flax fibre is proposed to reduce noise, vibration, and harshness throughout the structure. A study performed by the

Composites Evolution ("Reducing the Cost, Weight and NVH of Carbon Fibre," 2014) has found that a carbon/flax hybrid system is 15% cheaper, 7% lighter, and displays 58% greater vibration damping qualities over a full carbon fibre composite. Also, the flexural modulus is almost identical to carbon fibre, the latter scoring 47 GPa and a carbon/flax hybrid composite achieving 44 GPa. The company uses a 50/50 ratio of carbon/flax fibre, with the outermost layers consisting of carbon fibre.

A very interesting point has been made [4] that natural fibre composites offer an almost Carbon Dioxide (CO_2) neutral disposal process based on the captured CO_2 in natural fibres during their growth.

A growing awareness of industrial environmental impact has stimulated research into the development of environmentally friendly and sustainable materials [5]. Dhakal et al. investigated the effects of fibre orientation and thickness of natural fibres under an impact load. This study characterises the damage mechanisms in natural fibres throughout an impact event. It finds that after the samples are loaded beyond their elastic limit, damage begins to occur, in the form of matrix cracking. As the load continues the increase, the further onset of damage is seen as interfacial debonding as the specimen reach their peak loading. After this point, delamination and fibre breakage takes place until ultimately the sample is penetrated by the hemispherical tup. The orientation of fibres, fibre volume fraction, and matrix properties all have a significant effect on the damage type and severity observed.

Research into flax fibre reinforced epoxy composites [6] suggests that while flax may be considered one of the strongest natural fibre replacements for synthetic fibres, data on the transverse, shear, and compressive response of flax reinforced components is limited. The study found that delamination and fibre breakage is most prevalent in shear failure; while defibrillation and fibre cracking is presents under tensile loading. They suggest that matrix-related damage events, such as cracking and plasticity, are not a significant contributor to damage initiation or failure in flax composites.

The work undertaken by Sarasini et al. [7] studied the effects of layer sequencing on carbon/flax hybrid composites. An impact test in their work was carried out on four different configurations at energies between 5 and 30 J, in 5 J increments. While flax showed a better energy absorption capacity, it suffered greater internal damage and high compliance. The study found that the arrangement of carbon fibre on the outer layers, with inner flax fibre ply, has the best flexural performance. The damage pattern in the carbon samples showed a propagation of shear cracks moving far away from the impact zone, whereas the flax samples suffered heavy delamination. The samples with outer flax layers saw better mechanical and impact absorption properties over using a flax core. The flax samples began to show signs of penetration after 30 J, in 18-layer samples.

A study into natural fibre hybridisation by Dhakal et al. [8] looked into the performance of a hybrid natural fibre composite material, of hemp/basalt. The study found that natural fibres alone suffer critical issues with low post-impact residual damage tolerance through early fibre fracture and matrix cracking; however, the basalt skins assisted in delaying fractures of the hemp core, suggesting there are grounds for further investigating natural fibre hybridisation.

The effects of hybridising natural fibres with other materials [9], in this case, basalt, have brought an improvement of mechanical properties, such as improved resistance to impact damage and residual flexural strength properties compared to non-hybrid composites.

In recent years, critical engineering sectors, such as automotive, marine and aerospace are looking for lightweight composite materials to reduce their overall cost and weight with improved functionality [10]. The main goal of this study is to investigate the influence of carbon fibre hybridisation on the mechanical properties of carbon fibre epoxy, flax fibre epoxy, and a hybrid carbon/flax epoxy composite structure. This will be of direct benefit to industries aiming to reduce their carbon footprint by investigating a combination of natural and synthetic materials, which offer greater mechanical properties in certain applications.

Furthermore, using a variety of damage characterisation methods, this study will attempt to understand and highlight the failure mechanisms of hybrid systems, which will be useful for design engineers using composite materials to design components.

2. Materials and Methods

2.1. Materials

The two reinforcing materials used were epoxy-based prepregs 'HexPly M56' unidirectional carbon fibre and 'SDH VTC401LV' unidirectional flax fibre. Epoxy-based carbon and flax reinforcements used were obtained from Gurit and SHD Composites, respectively. The 'HexPly M56' [11] unidirectional carbon tape epoxy based prepreg, with a fabric weight of 280 g/m^2 supplied by Gurit, has a fibre density of 1.78 g/cm^3.

The flax fibre prepreg unidirectional mats with a fabric weight of 350 g/m^2 were obtained from SHD Composites, based on a VTC401 epoxy component. The flax fibres have a density of 1.5 g/cm^3, and in this case, the fibre volume of the prepreg is 50%.

2.2. Sample Preparation

The samples have the same layup procedure before being cured in the oven to their respective manufacturer specifications. The unidirectional prepreg is laid up into generic sheets of eight layers with a stacking sequence specified in Table 1. This ensures that the interface between carbon and flax in the hybrid composite is opposed at 90 °C and that there is a symmetrical distribution of fibre plies. The averages of fibre volume fraction (FVF) for flax/epoxy, carbon/epoxy, and flax-carbon/epoxy hybrid composites were approximately 56%, 59%, and 58%, respectively.

Table 1. Test specimen layup characteristics.

Specimen	Layers	Stacking Sequence (°)	Material Sequence
Flax/epoxy	8	0/+45/−45/90/90/−45/+45/0	F_8
Carbon/epoxy	8	0/+45/−45/90/90/−45/+45/0	C_8
Flax-carbon/epoxy hybrid	8	0/+45/−45/90/90/−45/+45/0	$C_2F_4C_2$

The material uses a vacuum bag to de-bulk and removes as much air as possible; a test is carried out by sealing the bag and removing the applied vacuum to ensure there are no vacuum leaks.

The samples were cured under similar conditions. The only difference was their ramping and dwelling temperatures, which were from 20 °C to 180 °C ± 5 °C and 180 °C ± 5 °C, respectively for CFRP composite sample, and 20 °C to 135 °C ± 5 °C and 135 °C ± 5 °C for FFRP and its hybrid samples. These temperatures were effective to obtain expected full curing. To ensure full cure of the matrix, a differential scanning calorimetry (DSC) test was performed and the correct glass transition temperature was measured.

Once the layup is complete, and the samples have been correctly de-bulked, the panels were placed in the oven for a controlled curing cycle as specified by the manufacturer of the prepreg epoxy resin. Temperature ramps are strictly controlled to ensure that the resin correctly cures; otherwise high-temperature snap curing can have reduced effectiveness as the impregnated resin is not allowed to flow to specification. After successful curing, the samples were CNC waterjet cut to sprue style templates for final collection and damage characterisation testing.

2.3. Low-Velocity Falling Weight Impact Testing

An impact test was undertaken on ZwickRoell HIT230F (ZwickRoell GmbH, Ulm, Germany), using preformed impact test samples. The incident impact energy was set at 25 joules (enough to penetrate the flax samples); with an impact velocity of 1.468 m/s and a total mass of 23.11 kg from a height of 110 mm. The specimens were firmly fixed at all edges using annular clamps with inner and outer diameters of 50 and 75 mm respectively. The specimens were cut by waterjet cutting from the laminate to a specimen size of 70 mm × 70 mm. Four specimens were impacted per each composite category and average values were taken.

The data obtained from the test was used to understand and evaluate the behaviour of carbon fibre alone, flax fibre alone, and carbon/flax hybrid composites under impact loading. It is important to understand how the material is deforming, and the failure modes that are present.

The impact samples were fully supported on a hardened steel retaining surface. Each specimen's thickness was measured in 90° incremental rotations using calibrated digital calipers. An average thickness, 2 mm for each sample, was obtained for each sample and then further averaged to give a total specimen thickness.

2.4. Flexural Testing

The flax/epoxy, carbon/epoxy and flax-carbon/epoxy hybrid composites were tested for determining flexural strength and modulus using a three-point bending test on a ZwickRoell Z030 (ZwickRoell GmbH, Ulm, Germany) machine in accordance with the BS EN 2746:1998 test method. A total of five samples were tested for each type of composite with a crosshead speed of 2 mm/min. The span-to-thickness ratio was kept at more than 16 times the thickness of the specimens. The panel thickness was approximately 2 mm for each specimen. Four specimens from each composite laminate were tested, and average values were taken.

The width and thickness of each sample were measured in three locations evenly distributed across the specimen's length. An average of the measurement data was obtained to be used to calculate the cross-sectional area, which was ultimately used to calculate the flexural strength and modulus of the specimen.

2.5. Damage Modes Characterisation

2.5.1. SEM

The fractured surfaces of failed samples under impact and flexural loadings were cut to fit within the vacuum chamber of the Zeiss Evo 10 scanning electron microscope (SEM) (Carl Zeiss Microscopy GmbH, Jena, Germany). The parted samples were then individually bagged to reduce contamination and then bonded to aluminium mounting stubs, and the specimen is coated in gold/palladium (Au/Pd) before entering the vacuum chamber.

2.5.2. Visual Inspection

The samples were catalogued with a digital camera; failure modes were observed and recorded.

3. Results and discussion

3.1. Impact Damage Characteristics

Three different types of composite materials were investigated in this study, namely: flax/epoxy, carbon/epoxy, and flax-carbon/epoxy hybrid composites. The impact test results, shown in Figure 1, are a comparison of these three composite types, calculated by taking the average for each material and finding the sample with the smallest deviation from the average.

In Figure 1a, it is noticeable that the plain flax/epoxy sample shows lower impact force during the impact event, with no return load (rebound) showing that the material has been completely penetrated with a lowest peak force of approximately 0.93 kN, and a highest deflection of approximately 12 mm. The rise in the displacement curve is consistent with the travel of the hemispherical tup impacting the flax specimen and then each layer taking up the slack, finally reaching the fracture point where the tup begins to traverse the topmost layer down consistently through each subsequent layer until it pierces the bottom-most layer.

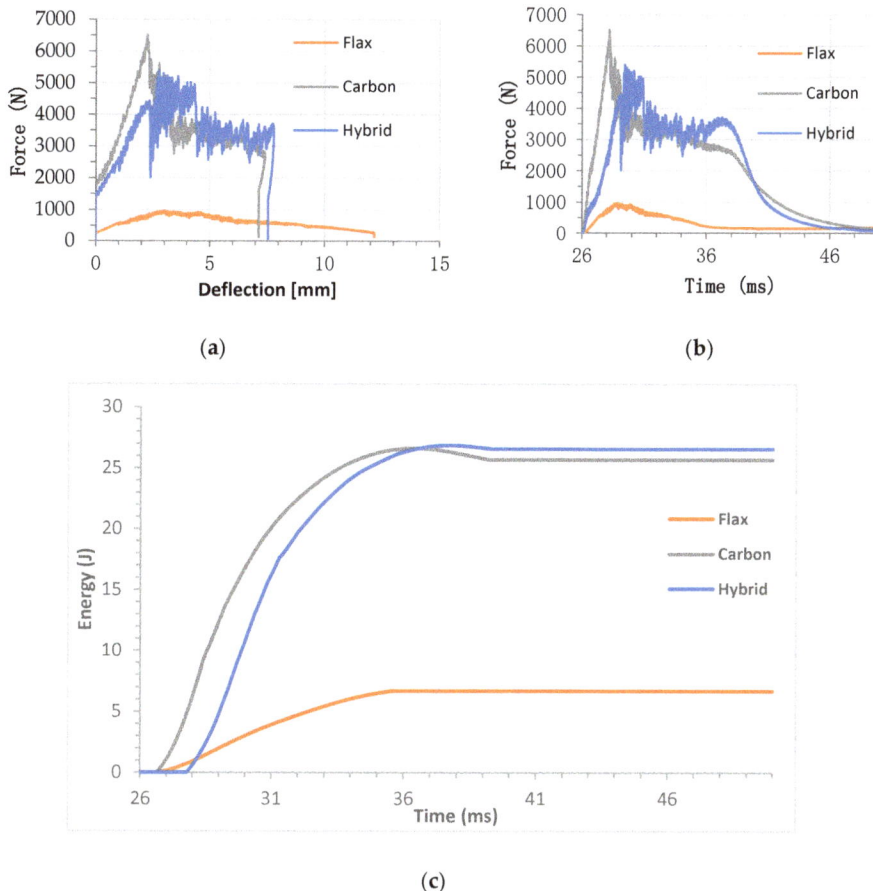

Figure 1. Impact test traces (**a**) force vs. deflection trace, (**b**) force vs. time trace, (**c**) energy vs. time.

The carbon/epoxy curve shows the highest impact force, approximately 6.51 kN, with the lowest deflection, approximately 7 mm. The rebound in the force-displacement curve is an indication that the impact probe has not sufficiently penetrated the sample.

The flax/epoxy sample had a greater deflection than the carbon/epoxy and flax-carbon/epoxy hybrid sample, but significantly lower impact force. However, the flax-carbon/epoxy hybrid specimen exhibited slightly higher deflection than the carbon/epoxy specimen with slightly lower impact force, approximately 5.39 kN. A point worthy of highlighting here is the deflection at peak force. The deflection recorded for the hybrid specimen is 2.74 mm, which is higher than that of the carbon/epoxy specimen. Similar observations can be made in peak energy. The flax/epoxy samples have shown the lowest energy absorption, approximately 7 joules, whereas the flax-carbon/epoxy hybrid sample had an almost identical energy (27 joules) to that of the carbon/epoxy samples shown in Figure 1c. This is due to higher damping properties of the flax core ply inside the hybrid composite. As the impact event is occurring and each layer takes up slack, the flax layers are able to absorb a greater amount of energy than that of the carbon fibre outer layers when they are put in tension. Because of this, the flax fibre inner layers will fail before the carbon fibre outer layers; experiencing debonding, delamination, and fibre pull-out before the failing of the carbon layers. This is shown in the trace for Figure 1a; as the load is applied and slack is taken up it moves at a constant rate, however after a deflection of 2 mm the carbon/flax specimen experiences an initial drop in force where the impact weight enters freefall. This is because the inner flax fibres delaminate from the carbon outer layer. Once the carbon layer takes up the slack again it cannot handle the shock load and begins to fail; after this point, the topmost carbon layers debond longitudinally to the unidirectional layup, with the carbon fibres finally breaking after 4 mm of drop weight travel through the sample. These observations can be related to the front and rear faces of the impacted samples.

The hybrid carbon-flax/epoxy does not reach peak load before serious fibre breakage, or delamination begins to occur in the data of Figure 1a. The force transferred into the impact sample drops momentarily by 2 kN. At the same time the work exerted on the sample has a small plateau at 28 ms into the rest as shown in Figure 1c. The force then climbs until reaching the peak load and oscillates as the impact tip tears through the fibres and matrix layers.

What is very interesting is how the carbon/flax hybrid sample shows harmonic resonance after the initial flax inner fibre failure [12], where the force applied also rings as it is dampened. Here, the flax layers, which have not yet failed, are damping the resonance which the carbon layers are experiencing.

The carbon/flax hybrid shows a similar pattern to carbon fibre with similar deformation potential; however, the downslope shows greater step sizes due to the different failure modes of the hybrid composite. The interfaces between the immediate carbon and flax layers proved to be weak and showed a very large delamination affected zone.

Another recent study supports the carbon/flax impact results [7], which shows a hybrid carbon/flax sample with a flax-fibre core exhibiting a peak force 82% below that of carbon; this report shows the hybrid carbon/flax sample demonstrating a peak force of 84.5% of carbon alone.

As the hemispherical impact tup traverses through the impact sample, plain carbon fibre epoxy and flax-fibre epoxy both exhibit predictable behaviour; however, the hybrid samples show interesting behaviour.

The carbon sample has a consistent application of force until it has reached its peak load at 2.237 mm. Between 2 mm and 4 mm of displacement, the impact object traverses the multiple layers of the sample, with a sharp reduction in force of 1500 N every 0.5 mm as it breaks a new layer until it comes to rest after breaking every fibre layer.

In Figure 1c, flax-fibre shows a smaller total amount of energy transferred between the probe and the sample, with the rate of transfer having a slower curve than that of the other samples. The probe comes to rest after penetrating the sample approximately 10 ms into the test, with force ceasing to be applied once maximum deformation has been reached. This is due to the difference in the stresses between the flax fibres and the matrix interface being large enough for debonding and delamination to begin to occur earlier than in the plain carbon or carbon/flax samples [2].

The carbon/epoxy and flax-carbon-epoxy hybrid systems show a consistent downslope in Figure 1b after 36 ms, due to energy being transferred back into the impact probe, as the fibres (still within their elastic limit) return to their original elongation. The carbon/epoxy specimen shows a more consistent reduction in the force applied until recoil; however, flax exhibits an arc of force applied to increase before recoil, demonstrating the dampening properties of the flax layers within the sample. Similar positive hybrid effects on the impact behaviour of natural fibre composites were reported by Sarasini et al. [13]. with intraply hybrid flax-basalt composites. The natural fibre reinforced composites have low impact resistance behaviour compared to their conventional counterparts, such as glass and carbon fibre reinforced composites. A significant impact properties enhancement with the carbon fibre hybridisation is a very positive achievement towards using these sustainable composites as an alternative to pure synthetic composites in load-bearing applications while maintaining their partial green attributes.

3.2. Flexural Properties

The average flexural properties of three different types of composites are presented in Table 2, and load vs. deformation traces of these composites are shown in Figure 2. It can be extrapolated from the results illustrated in Figure 2 that flax-carbon/epoxy hybridised samples have shown a significant improvement in flexural strength and modulus. Precisely, the flexural strength of plain flax/epoxy increases significantly from 95.66 MPa to 425.87 MPa (an approximate 345% improvement) with carbon fibre hybridisation. Similarly, the flexural modulus of plain flax composite was increased from 4.78 GPa to 17.90 GPa (an approximate 274% improvement) with carbon fibre hybridisation. These values represent the highest mean value amongst the studied composites. The significant enhancement in flexural modulus is dependent on several factors such as fibre content and modulus of fibre itself. Moreover, the compatibility between flax and carbon fibre as well as matrix and reinforcements may have contributed to the improvement in flexural modulus. This improvement is further attributed to the effect of hybrid mechanisms. The lay-up sequence for hybrid composites was two layers of high-modulus carbon fibres on the outside surfaces, and the pure flax fibre in the middle has contributed the highest strength and modulus. It is worth noting that flax fibre is a very stiff material which has further contributed to this significant flexural properties' improvement. The attainment of such property enhancement with carbon fibre hybridisation provides a significant potential of natural fibre hybrid composites to be used for structural light weight applications [14].

Table 2. Average flexural properties obtained from three-point bending testing.

Specimen	Peak Force (N)	Flexural Strength (MPa)	Flexural Modulus (GPa)	Deformation at Peak Force (mm)
Flax/epoxy	115.75 (±6.61)	95.66 (±5.46)	4.78 (±1.16)	4.01 (±0.32)
Flax-carbon/epoxy hybrid	553.30 (±61.63)	425.87 (±50.93)	17.90 (±0.31)	3.96 (±0.22)
Carbon/epoxy	532.40 (±9.55)	464.65 (±7.89)	52.82 (±2.16)	1.16 (±0.13)

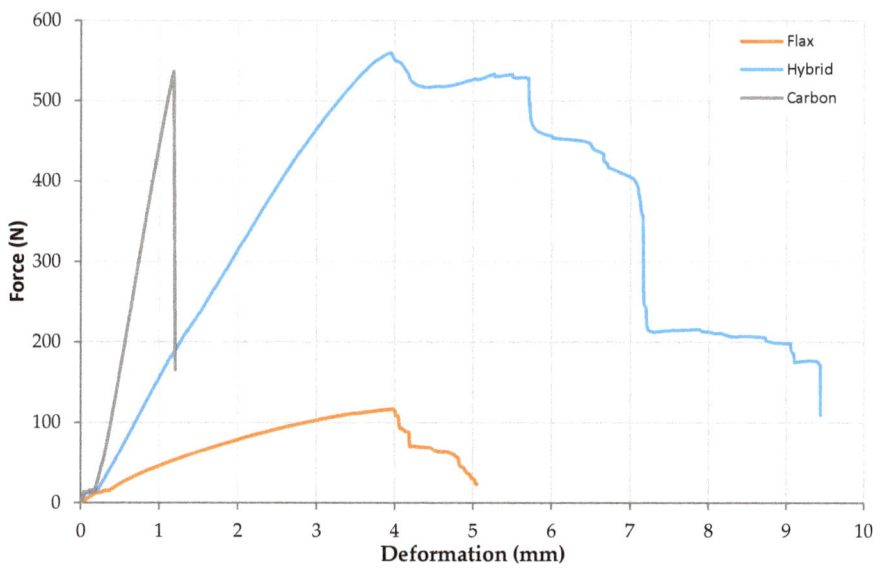

Figure 2. Force versus deformation traces obtained from flexural testing of flax/epoxy, carbon/epoxy and flax-carbon/epoxy hybrid composites.

The incredibly high-flexural properties of the carbon/flax hybrid could support a theory of a very strong interfacial relationship between carbon fibre and flax fibre in an epoxy laminate under flexural load. Similarly, the flexural deformation was significantly higher, increasing from 1.16 mm to 3.96 mm (an approximate 241% improvement) for flax-carbon hybrid systems compared to carbon/epoxy systems, indicating a hybrid system is a valid approach towards achieving an improved mechanical performance of natural fibre reinforced composites.

3.3. Damage Characterisation

3.3.1. SEM Images of Plain Flax/Epoxy Composites under Impact

Scanning electron microscopy (SEM) images of fractured surfaces after the impact of plain flax/epoxy composites are presented in Figure 3a,b which shows extensive fibre breakage and disorder, with one large group of fibres becoming an initial focal point. The following magnification scales (150 and 300), display matrix cracking and debonding of the epoxy from individual fibres, and additionally show the fibre bending and debonding around a kink band of the flax fibres structure, with clear twisted and flattened fibres. Similar failure mode under the low velocity impact testing was reported by Dhakal et al. for hemp fibre reinforced unsaturated based composites [15].

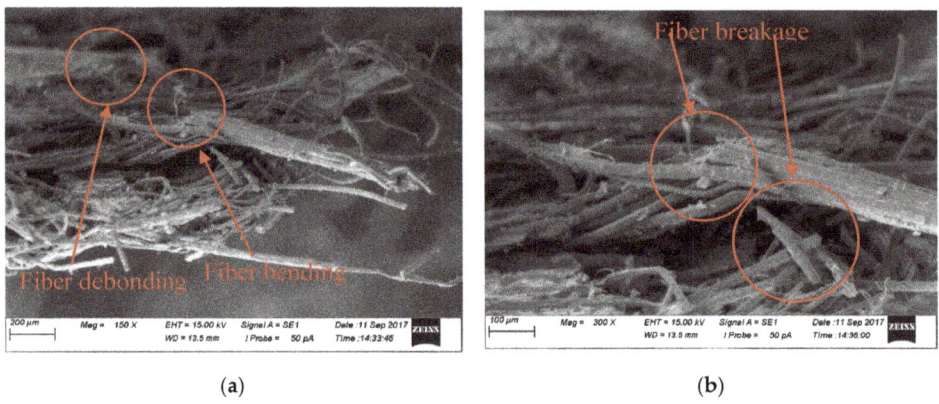

Figure 3. SEM images of fracture surface morphology of plain flax composites failed under impact loading at different magnifications (**a**) fibre debonding and bending; (**b**) fibre breakage.

3.3.2. SEM Images of Plain Flax/Epoxy Composite under Flexural Loading

SEM images of the fractured surfaces of plain flax/epoxy composites following flexural loading are shown in Figure 4. In Figure 4a, the tensile (T) and compressive (C) load paths have been annotated. It is clear that under three-point bending, natural fibres are heavily affected by not only the tensile stresses but also compression which causes a large amount of compaction on the bottom of the image, where the loading nose would exert force. This could cause excessive debonding and shear slippage.

Figure 4b shows the result of the fractured surface after the flexural test, as the outer layer has been debonded from the inner layers at a 0/+45° intersection of the flax fibre epoxy, with a large portion of the epoxy matrix released, shown in Figure 4c, from the crack with several fibres still attached. More enhanced views in Figure 4c show the origin of the released matrix bundle, with highly fragmented matrix portions at this site.

3.3.3. SEM Images of Plain Carbon/Epoxy Composites under Impact Loading

Figure 5 shows uniform breakage as an outer layer of flax fractures upon receiving a flexural load transverse to the plane of the unidirectional fibre layer. This perspective would be facing the impact tup as it travels through the SEM image.

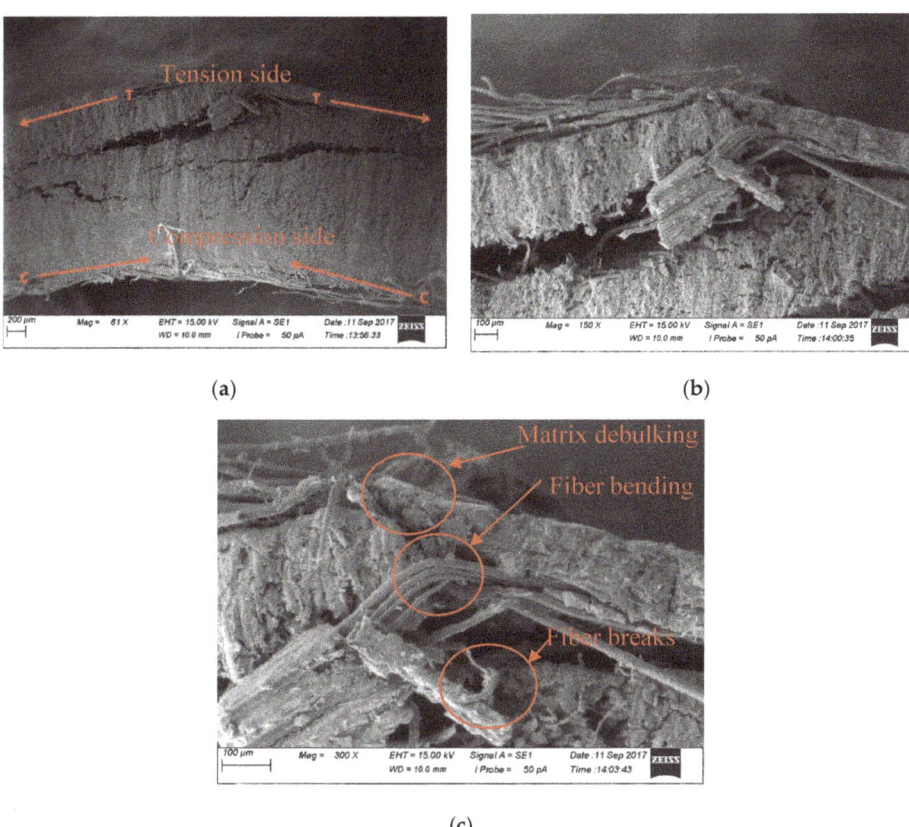

Figure 4. SEM images of fracture surfaces of flax alone composites failed under flexural loading (**a**) showing tension and compressive load path; (**b**) debonding and large part of matrix debulked; (**c**) fibre bending, fibre breaks and matrix debulking shown at both compression and tension sides.

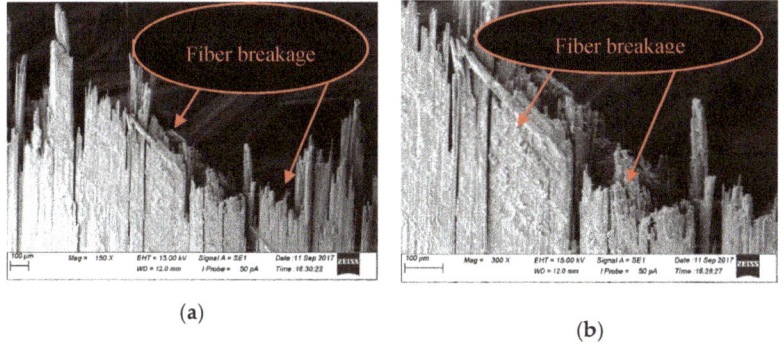

Figure 5. SEM images of the outermost layer of carbon fibre impact samples (**a**) fibre breakage at 150× magnification; (**b**) fibre breakage site at an enhanced 300× magnification.

Figure 5a shows that along with uniform fracture points on each fibre, the severe delamination pattern from the released outermost layer of carbon still presents in the epoxy matrix. This pattern is more visible in Figure 5b; the fibres are failing with small delamination visibly occurring and cracks running longitudinally along fibres.

Moving the focal length down into the impact sample in Figure 5, the multiple layers are visible with excessive breakage apparent as fibres have been pushed between layers by the impact event.

3.3.4. SEM Images of Plain Carbon/Epoxy Composites under Flexural Loading

The image above, Figure 6a,b, shows multiple carbon fibre layers after having performed a three-point flexural test. Figure 6 shows a very uniform pattern of fibres, even after breakage. This is highly contrasting to the flax fibre failures, where individual fibres are chaotic, such as flax in Figure 4. Figure 6 shows fibre breakage, and a cluster of fractures fibres 100 µm long are distorted away from the layer's plane, however the fibres continue to orient themselves through a cohesive matrix; this is a good example of the brittle nature of carbon fibre epoxy. It can be seen that the matrix is seen to be partially fragmented, but as the failure was not an explosive release of high tension the matrix did not shatter and spread itself as an airborne particulate across the surrounding material.

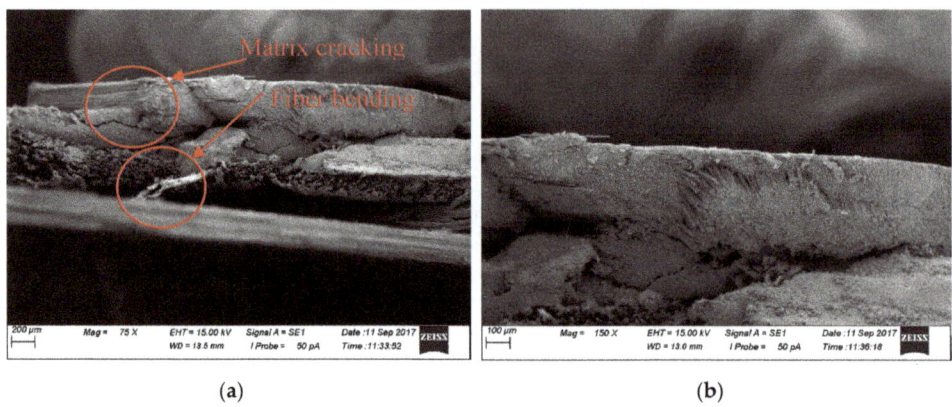

Figure 6. Plain carbon/epoxy composites under 3-point bending testing sample at different magnifications (**a**) matrix cracking and fibre bending at 75× magnification; (**b**) matrix cracking and fibre bending at an enhanced 150× magnification.

3.3.5. SEM Images of Flax/Carbon/Epoxy Hybrid Composites under Impact

Figure 7a,b display the internal flax fibre core from the hybrid composite impact test, which exhibits a wide area of damage through the fibres, with a high degree of fibrillation present on the large fibres. When enhancing Figure 7c, the portion of the matrix shows significant cracking, tearing and delamination; with the fibres still attached becoming highly twisted and distorted as the matrix has been ejected from the primary sample body.

A further view of Figure 7a,b display both carbon and flax fibres in a chaotic site, where flax fibres from the core piles have intersected the carbon layer. The carbon to flax interface is at a 90° orientation, at a +45° offset to the topmost layer of the sample. At a higher magnification, the expected result is seen from an explosive event in a carbon fibre matrix structure. The carbon fibres that have debonded from the epoxy matrix have scattered and have not remained as a group; such as with carbon three-point bending.

Furthermore, the third site of interest is a fantastic example of fibre pull-out and the matrix surface after complete delamination. Again, in Figure 7, it can be seen that the flax fibre epoxy matrix has delaminated from the carbon fibre interface, leaving a strong pattern of convex valleys at a 90° orientation to the flax fibre direction. After increasing the magnification, Figure 7c shows the same explosive carbon fibre matrix contamination left behind on the matrix surface of the flax fibre layer.

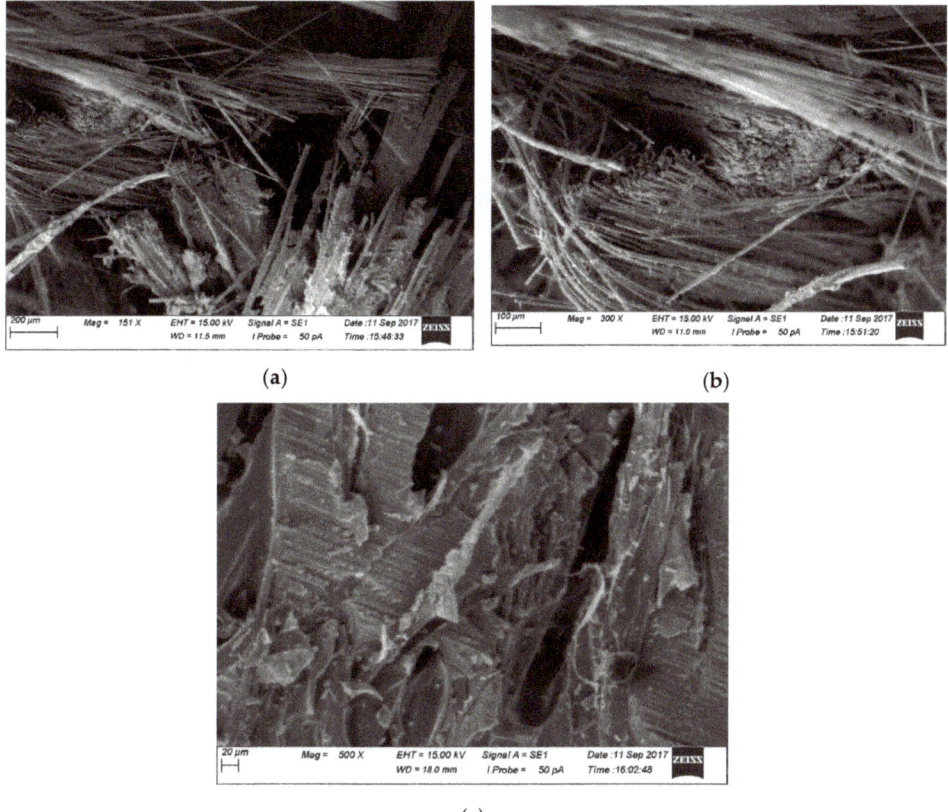

Figure 7. SEM images of flax/carbon/epoxy hybrid composites surfaces failed under impact loading at different magnifications. (**a**) flax fibres intersecting a carbon fibre layer at 151× magnification; (**b**) enhanced view of flax fibres intersecting a carbon fibre layer at 300× magnification; (**c**) fibre pull-out.

3.3.6. SEM Images of Flax-Carbon/Epoxy Hybrid Composites under Flexural Loading

Figure 8 shows an SEM micrograph of a hybrid carbon/flax/epoxy sample after the three-point bending test at different magnifications. The image has been annotated with the tensile (T) and compressive (C) load paths (Figure 8a).

Figure 8a further shows the interface between carbon and carbon (0°/+45°) showing complete delamination. Viewing the sample shown in Figure 8, there is a bundle of fibres delaminated from the main ply, this has caused the load to be transferred into the next later, which is flax.

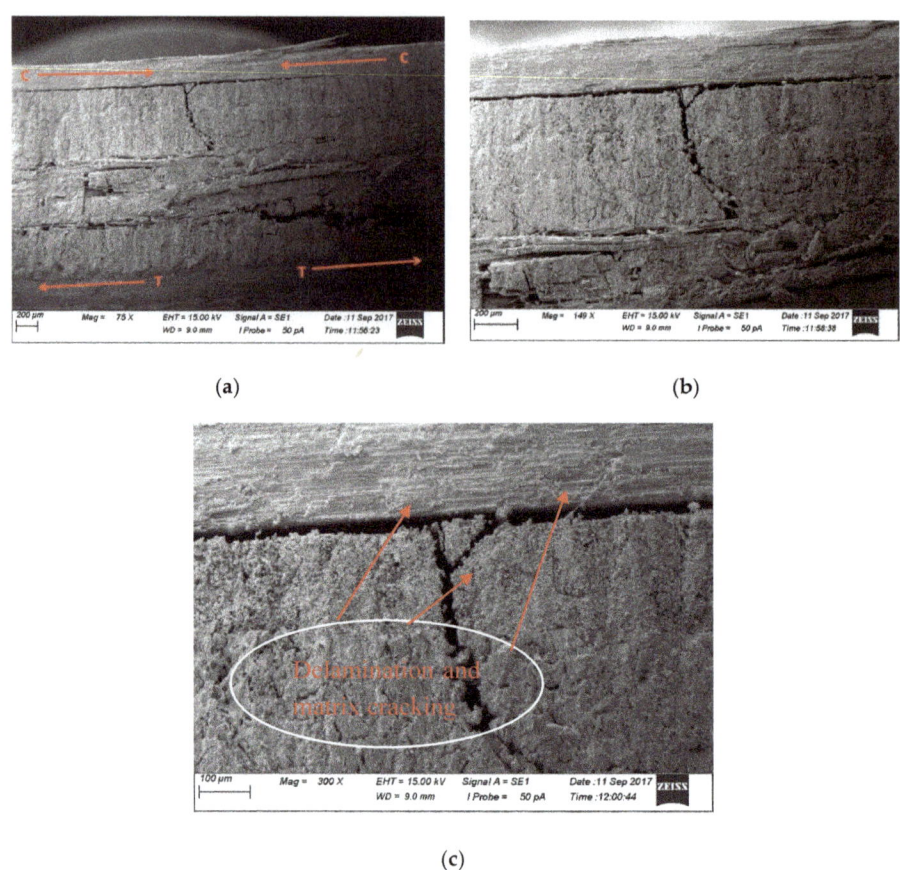

Figure 8. SEM images of flax/carbon/epoxy hybrid composites at different magnifications showing (**a**) tension and compression pathways, (**b**) delamination between carbon and flax fibres, (**c**) delamination and matrix cracking.

3.4. Visual Inspection of Damage Modes

3.4.1. Visual Observation of Impact Damage

Figure 9 displays the failed surfaces of impact loading showing rear sides of the samples, namely, flax/epoxy, carbon/epoxy and carbon-flax/epoxy hybrid structures.

The plain flax fibre epoxy impact samples showed complete drop weight penetration in all samples, shown in Figure 10a,b. Combined large cracks can be seen on the reverse images; the samples were more likely to suffer radial fractures upon penetration [13] (pp. 559–567). The damage is within an area similar to the diameter of the impact tup.

The topside of the flax fibre impact samples experiences fibre breakage spanning the circumference of the impact zone, shown in Figure 10a. The impact tup then folded the topmost layers through the penetration zone.

The carbon impact samples shown in Figure 10c,d only saw bent fibres around the impact dent as the tup did not penetrate. The tup did split the upper layer, but no signs of large delamination are present on the topmost layer.

However, the bottommost layers of the carbon fibre impact samples saw push out delamination, shown in Figure 10c,d. The rearmost layer has delaminated in 2 mm wide splinters, rising in an opposing pattern over the rear of the impact site, in some cases the first layer behind the outermost rear layer has seen fibres protrude beyond the eighth layer.

The hybrid samples, pictured in Figure 10e,f, show the impact site with a wide area of damage suspected to be the flax fibres crushing and the matrix cracking heavily internally leading to delamination from the carbon. As with the carbon only samples, there is excessive push out delamination on the rearmost layer, with a large amount of fibre breakage and debonding.

Figure 9. Post-impact test underside damage (**a**) flax fibre epoxy, (**b**) carbon fibre epoxy, (**c**) carbon-flax/epoxy hybrid composites.

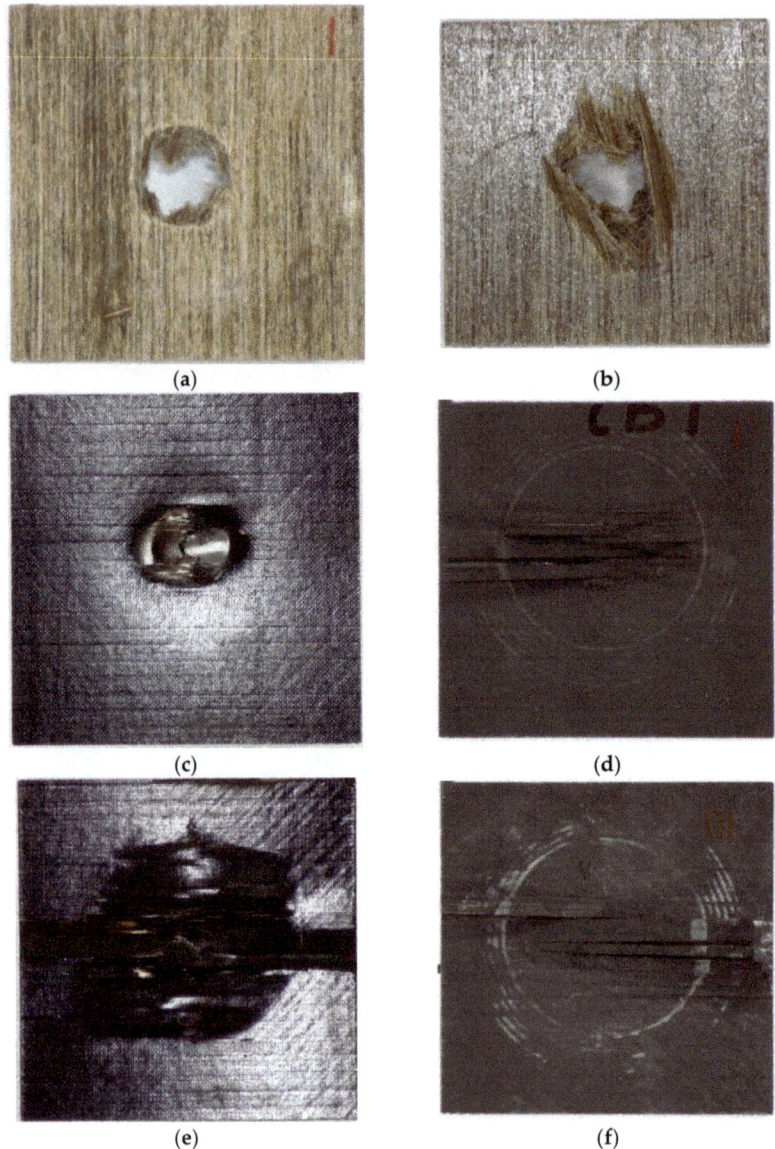

Figure 10. Damage on impacted front and rear surface of flax, carbon and flax/carbon epoxy hybrid composites (**a**) flax front surface, (**b**) flax rear surface, (**c**) carbon front and (**d**) carbon rear, (**e**) carbon/flax hybrid front, (**f**) carbon/flax hybrid rear surface.

3.4.2. Visual Observations of Damage under Flexural Loading

The flax/epoxy samples all displayed crushed fibres around the point of the loading nose contact, where the fibres are under compression. This is shown in Figure 11a above with arrows marking the compressive and tensile load paths, including Figure 11b. The compressed topmost layers see excessive random delamination and matrix cracking. The bottom layers under tension see delamination and more fibre breakage.

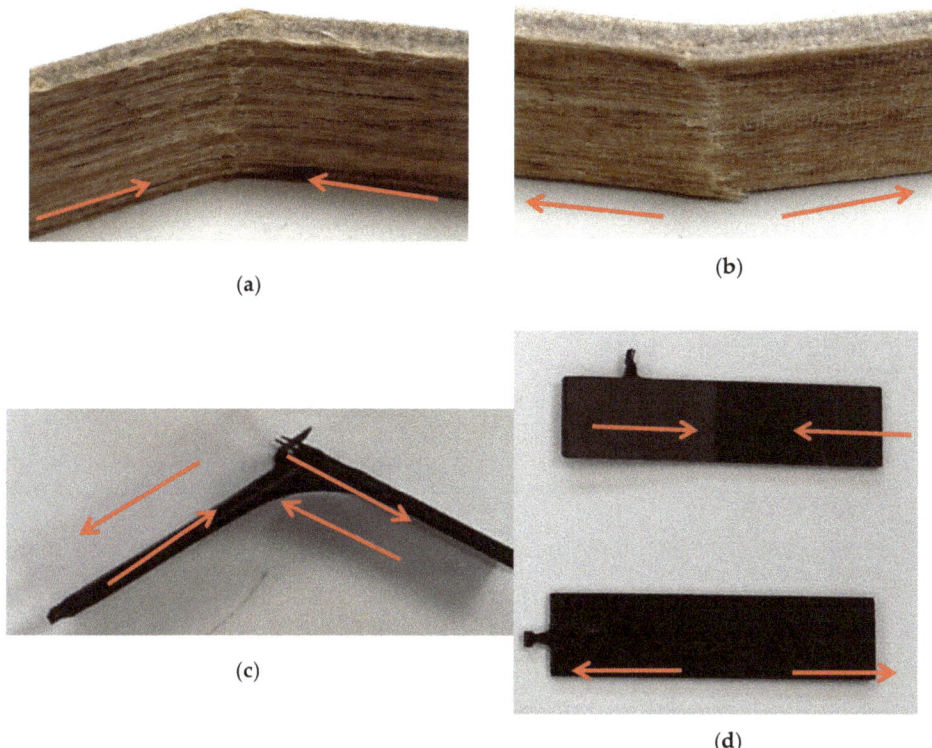

Figure 11. Failed samples under flexural loading (**a**) compressed plain flax/epoxy, (**b**) split outer layer of plain flax/epoxy under tension, (**c**) plain carbon/epoxy damage, (**d**) flax-carbon/epoxy hybrid composites.

The topmost layer of carbon fibre under tension, shown in Figure 11c, suffers massive delamination from the primary body, which begins to experience a greater form of fibre breakage and debonding until the primary layers under tension are shown to have an explosive failure. The hybrid samples in Figure 11d show minimal external damage visible using non-destructive testing methods.

4. Conclusions

The mechanical properties (low-velocity impact and flexural) of flax/epoxy, carbon/epoxy and flax-carbon/epoxy hybrid composites were experimentally studied. This study has clearly suggested that carbon fibre hybridisation onto flax/epoxy composites can contribute a significant improvement in impact damage behaviour and flexural strength and modulus.

Through damage analysis, the hybrid composite displayed similar impact characteristics to the plain carbon/epoxy composites, far exceeding the performance of plain flax/epoxy composites alone. It was also evidenced that the natural fibres such as flax also dampened the harmonic resonance during the test. This is a significant achievement in providing the potential of natural fibre hybrid composites in semi-structural and structural light-weight applications.

The damage characterisation through SEM imaging has shown the various failure modes of the plain flax/epoxy and flax-carbon/epoxy hybrid composites, such as the shock loading of flax fibre cores in falling weight impact and flexural loading scenarios.

Author Contributions: M.C. contributed in design, conducting experimental procedures and compiling the results as well as writing the preliminary report. H.N.D. provided the overall guidance (supervision) including the design of the study, general interpretation of the results and writing of the paper.

Funding: This research received no external funding.

Conflicts of Interest: The authors declare no conflict of interest.

References

1. European Comission. Available online: https://ec.europa.eu/clima/policies/transport/vehicles/cars_en (accessed on 25 March 2019).
2. Dhakal, H.; Zhang, Z.; Guthrie, R.; MacMullen, J.; Bennett, N. Development of flax/carbon fibre hybrid composites for enhanced properties. *Carbohydr. Polym.* **2013**, *96*, 1–8. [CrossRef] [PubMed]
3. Composites Evolution. Available online: https://compositesevolution.com/news/carbonflax-hybrid-automotive-door-on-display-at-jec-world-featuring-composites-evolutions-biotex-flax/ (accessed on 25 March 2019).
4. Pil, L.; Bensadoun, F.; Pariset, J.; Verpoest, I. Why are designers fascinated by flax and help fibre composites? *Compos. Part A Appl. Sci. Manuf.* **2016**, *83*, 193–205. [CrossRef]
5. Dhakal, H.N.; Skrifvars, M.M.; Adekunle, K.; Zhang, Z.Y. Falling weight impact response of jute/methacrylated soybean oil bio-composites under low velocity impact loading. *Compos. Sci. Technol.* **2014**, *92*, 134–141. [CrossRef]
6. Mahboob, Z.; Sawi, I.E.; Zdero, R.; Fawaz, Z.; Bougherara, H. Tensile and compressive damaged response in flax fibre reinforced epoxy composites. *Compos. Part A Appl. Sci. Manuf.* **2017**, *92*, 118–133. [CrossRef]
7. Sarasini, F.; Tirillò, J.; D'Altilia, S.; Valente, T.; Santulli, C.; Touchard, F.; Chocinski-Arnault, L.; Mellier, D.; Lampani, L.; Gaudenzi, P. Damage tolerance of carbon/flax hybrid composites subjected to low velocity impact. *Compos. Part B Eng.* **2016**, *91*, 144–153. [CrossRef]
8. Dhakal, H.N.; Sarasini, F.; Santulli, C.; Trillò, J.; Zhang, Z. Effect of basalt fibre hybridisation on post-impact mechanical behaviour of hemp fibre reinforced composites. *Compos. Part A Appl. Sci. Manuf.* **2015**, *75*, 54–67. [CrossRef]
9. Kumar, S.C.; Arumugam, V.; Dhakal, H.N.; John, R.R. Effect of temperature and hybridisation on the low velocity impact behaviour of hemp-basalt/epoxy composites. *Compos. Struct.* **2015**, *125*, 407–416. [CrossRef]
10. Woigk, W.; Fuentesc, C.A.; Riond, J.; Hegemanne, D.; van Vuurec, A.W.; Dransfeldb, C.; Masania, K. Interface properties and their effect on the mechanical performance of flax fibre thermoplastic composites. *Compos. Part A Appl. Sci. Manuf.* **2019**, *122*, 8–17. [CrossRef]
11. Hexcel. Available online: http://www.hexcel.com/Resources/DataSheets/Prepreg (accessed on 25 March 2019).
12. ASTM International. *Standard Test Method for Measuring the Damage Resistance of a Fiber-Reinforced Polymer Mareix Composite to a Drop-Weight Impact Event*; ASTM International: West Conshohocken PA, USA, 2012.
13. Sarasini, F.; Tirillo, J.; Ferrante, L.; Sergi, C.; Russo, P.; Simeoli, G.; Cimino, F.; Rosaria Ricciardi, M.; Antonucci, V. Quasi-static and low-velocity impact behaviour of intraply hybrid flax/basalt composites. *Fibers* **2019**, *7*, 26. [CrossRef]
14. Russo, P.; Simeoli, G.; Vitiello, L.; Fillippone, G. Bio-polyamide 11 hybrid composites reinforced with basalt/flax interwoven fibres: A tough green composite for semi-structural applications. *Fibers* **2019**, *7*, 41. [CrossRef]
15. Dhakal, H.N.; Zhang, Z.Y.; Richardson, M.O.; Errajhi, O.A. The low velocity impact response of non-woven hemp fibre reinforced unsaturated polyester composites. *Compos. Struct.* **2007**, *71*, 559–567. [CrossRef]

© 2019 by the authors. Licensee MDPI, Basel, Switzerland. This article is an open access article distributed under the terms and conditions of the Creative Commons Attribution (CC BY) license (http://creativecommons.org/licenses/by/4.0/).

Article

Tensile Behavior of Unidirectional Bamboo/Coir Fiber Hybrid Composites

Le Quan Ngoc Tran [1,*], Carlos Fuentes [2], Ignace Verpoest [2] and Aart Willem Van Vuure [2,*]

1. Agency for Science, Technology and Research, Singapore Institute of Manufacturing Technology, (A*STAR), 73 Nanyang Drive, Singapore 637662, Singapore
2. Department of Materials Engineering (MTM), KU Leuven, Kasteelpark Arenberg 44, 3001 Leuven, Belgium
* Correspondence: tranlqn@simtech.a-star.edu.sg (L.Q.N.T); aartwillem.vanvuure@kuleuven.be (A.W.V.V.); Tel.: +65-6793-8958 (L.Q.N.T.); +32-16-30 1138 (A.W.V.V.)

Received: 14 May 2019; Accepted: 6 July 2019; Published: 10 July 2019

Abstract: Natural fibers, such as bamboo, flax, hemp, and coir, are usually different in terms of microstructure and chemical composition. The mechanical properties of natural fibers strongly depend on the organization of cell walls and the cellulose micro-fibril angle in the dominant cell wall layers. Bamboo, flax, and hemp are known for high strength and stiffness, while coir has high elongation to failure. Based on the unique properties of the fibers, fiber hybridization is expected to combine the advantages of different natural fibers for composite applications. In this paper, a study on bamboo/coir fiber hybrid composites was carried out to investigate the hybrid effect of tough coir fibers and brittle bamboo fibers in the composites. The tensile behavior of unidirectional composites of bamboo fibers, coir fibers, and hybrid bamboo/coir fibers with a thermoplastic matrix was studied. The correlation between the tensile properties of the fibers and of the hybrid composites was analyzed to understand the hybrid effects. In addition, the failure mode and fracture morphology of the hybrid composites were examined. The results suggested that, with a low bamboo fiber fraction, a positive hybrid effect with an increase of composite strain to failure was obtained, which can be attributed to the high strain to failure of the coir fibers; the bamboo fibers provided high stiffness and strength to the composites.

Keywords: natural fibers; hybridization; Unidirectional (UD) composites

1. Introduction

Fiber-reinforced polymer composites are attractive materials for a wide range of applications due to their high strength and stiffness in combination with light weight. Modern composite structures are increasingly subjected to multiple performance criteria, in which the optimum combination of mechanical properties (e.g., strength, stiffness, and toughness), cost, and sustainability are considered. Fiber hybridization has recently received a high interest in research and application for creating hybrid composites having synergetic properties. The fiber-hybrid composites provide more design freedom than non-hybrid composites, and possibly lead to synergetic effects that neither of constituents possess [1–5].

In order to achieve the synergies in fiber-hybrid composites, several aspects play important roles, including selection of suitable fibers, selection of suitable fiber combination and understanding fiber interactions in the hybrid systems. While there is substantial information on fiber-hybridization for synthetic fiber composites [4,6–8], the understanding of hybrid effects in natural fiber composites is still limited.

Natural fibers, such as bamboo, flax, hemp, and coir, extracted from plants, are usually different in terms of microstructure and chemical composition. The mechanical properties of natural fibers strongly depend on the organization of cell walls and the cellulose micro-fibril angles in the cell wall layers. Bamboo, flax, and hemp are known for high strength and stiffness, while coir has high strain to failure [9,10]. Based on the unique properties of the fibers, fiber hybridization is expected to combine the advantages of the different natural fibers for composite applications.

In this study, tensile behavior of unidirectional composites of coir/bamboo polypropylene composites was characterized at the macro level, where fibers are mixed at the fiber layer scale. Thin coir and bamboo prepress (a thickness of 1–3 technical fibers) were used for making the hybrid composite samples with the intention of approaching a good mixing at the single fiber level, which is considered hybridization at the micro scale; theoretical studies [11,12] predict a better stress transfer in hybrid composites when the fibers are mixed at the micro level. The correlation between the tensile properties of the fibers and of the hybrid composites was analyzed. In addition, the failure mode and fracture morphology of the hybrid composites were also examined to provide a better understanding of the hybrid effects.

2. Materials and Methods

2.1. Materials

2.1.1. Fibers and Polymer Matrix

Technical coir and bamboo fibers were used in this study. The coir fibers were long coir with fiber length in the range of 200–300 mm, supplied by the Can Tho University of Vietnam, where the fibers were extracted from the husk shell of coconut from the coconut palm (*Cocos nucifera* L.). The technical bamboo fibers were extracted from bamboo culms of the species *Guadua angustifolia* (from Colombia), using a novel mechanical extraction process developed by KU Leuven, giving a maximum fiber length between 200 and 350 mm. The extracted coir and bamboo fibers were soaked in hot distilled water at 70 °C for 2 h, and then smoothly washed with alcohol to remove greases which may attach on the fiber surface during the fiber extraction process, rinsed with deionized water, and dried under vacuum at 90 °C.

Polypropylene (PP) was used as the matrix for composites. The PP was an unmodified grade and supplied in sheet form by Propex GmbH & Co. KG (Gronau, Germany). The polymer has melt flow rate of 5.2 g/10 min and melting temperature of 160.6 °C. The mechanical properties of the PP were measured by three point bending test following ASTM 790M, which are presented in Table 1.

2.1.2. Mechanical Properties

Tensile properties of the technical fibers were investigated in previous studies [9,10,13–15]. Table 1 shows the tensile strength, E-modulus, and strain at failure of the studied coir and bamboo fibers. The mechanical properties of PP were also assessed and presented.

Table 1. Mechanical properties of studied fibers and polymer matrix.

Material	E-Modulus (GPa)	Strength (MPa)	Strain to Failure (%)	Density (g/cm^3)	Reference
Coir fiber	4.6–4.9	210–250	18.0–36.7	1.3	[10]
Bamboo	42–50	775–860	1.1.9	1.4	[15,16]
PP	1.6–1.8	55–65	>300	0.9	tested values

2.2. Fabrication of Unidirectional (UD) Hybrid Composites

2.2.1. Preparation of UD Coir and Bamboo Prepreg with Polypropylene

The extracted coir and bamboo fibers were delivered in a bundle and slightly twisted. In order to make UD hybrid composites, it is required that the fibers are properly aligned in one direction. In this work, a procedure for fiber alignment was developed, in which the coir and bamboo fibers were soaked in water, then combed and evenly spread in a thin layer of UD fibers (with thickness of 1–3 technical fibers). This wet layer was placed between two plastic plates to keep the UD form of the fiber layer, during drying at 70 °C for three days in an oven. After drying, the UD fiber layers were used for making prepregs with PP matrix.

The prepreg was made by placing a UD fiber layer sandwiched between two layers of thermoplastic films, as seen in (Figure 1). The sandwich was clamped and pressed at approximately 200 °C by an iron in order to consolidate the fiber and the matrix to form a prepreg.

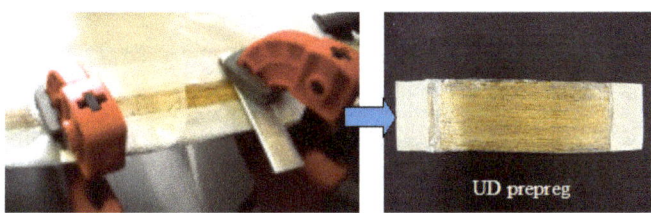

Figure 1. Preparation of UD coir fiber polypropylene prepreg.

2.2.2. Composite Processing

UD composites of coir and bamboo fiber with PP matrix were produced using prepregs in order to perform tensile tests. The test samples had dimensions of 15 mm × 250 mm × 2 mm (width × length × thickness) following ASTM 3039.

For composites processing, the prepregs were cut into the desired dimensions fitting into an Aluminum mold (Figure 2a) with designated stacking sequences. The thickness of the samples was controlled by placing aluminum stoppers at both edges of the mold channels between the upper and lower mold. Six samples of each type could be produced at one time using six channels in the molds. The fiber volume fraction of the composite samples was estimated by the weight of the fibers and the matrix films. Three types of UD composites were produced, including monolithic UD coir/PP, UD bamboo/PP, and UD coir-bamboo/PP composites. The closed mold set-ups were then placed into the Pinette hot press (Figure 2b) for composites fabrication, under processing parameters of 175 °C, at 10 bar pressure and for 15 min, after that the mold was cooled to room temperature under the same pressure.

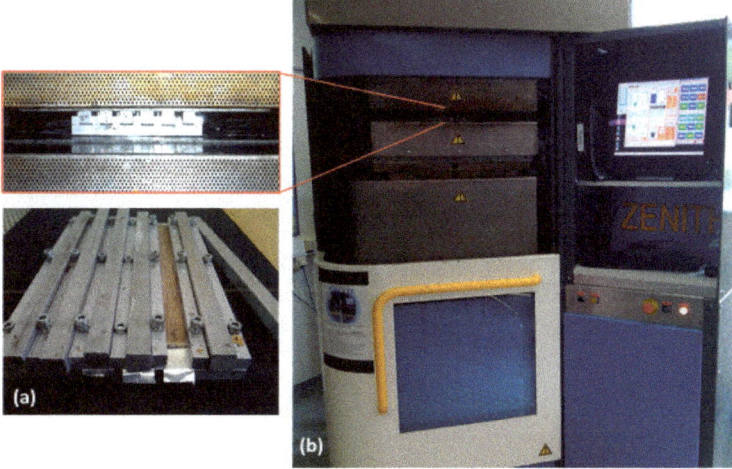

Figure 2. (a) Mold for tensile test samples, and (b) Composite processing in the Pinette hot press.

The hybrid coir-bamboo/PP composite samples were prepared by stacking coir/PP and bamboo/PP prepregs in a sequence of two layers of coir/PP prepreg at the outside and one layer of bamboo/PP prepreg in the middle. For monolithic composites, six layers of coir/PP and bamboo/PP prepreg were used for producing the composites.

The produced UD coir-bamboo hybrid composites had fiber volume fractions of coir and bamboo fibers of approximately 30% and 8%, respectively, while the fiber volume fraction of the coir/PP and the bamboo/PP is 44% and 45%, respectively.

2.3. Tensile Test and Characterization of Composite Microstructure

Tensile tests were performed according to the standard ASTM D3039, on composite samples of 15 mm × 200 mm × 2 mm, to which composite end-tabs were glued. A load cell of 5 kN was used and a crosshead speed of 1 mm/min was applied. The gauge length between the two clamps was set at 100 mm, while an extensometer with a gauge length of 50 mm was employed for measuring the sample strain. (Figure 3) shows the set up for the tensile test and some tested samples.

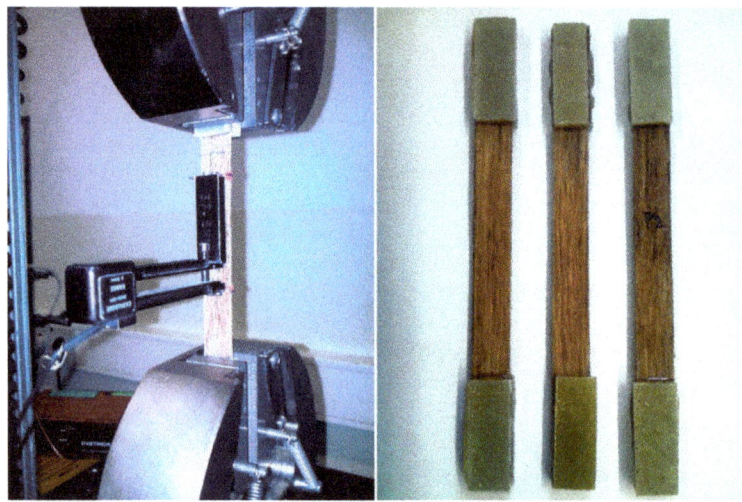

Figure 3. Tensile test (**left**) and test samples (**right**).

Various composite systems, including bamboo, coir, and hybrid coir-bamboo in PP, were characterized. Six samples for each type of composite were tested.

Scanning electronic microscopy (SEM) images of the composite cross-sections were taken after failure using a Philips XL 30 FEG scanning electron microscope (FEI Europe B.V., Zaventem, Belgium). The images provide the information of the fracture of the composites including fiber distribution and failure mechanism of the fibers.

3. Results and Discussion

3.1. Tensile Behavior of the Monolithic and Hybrid Composite

The tensile stress-strain curves of the UD coir-bamboo hybrid composites are presented in Figure 4. It can be seen that the hybrid composites show an almost linear-elastic behavior until a peak stress, and then the stress dramatically decreases to a certain value. From this point on, the stress reduces slowly in a plastic manner. From this behavior, it is suggested that the coir fibers and the bamboo fibers together carry the tensile load until reaching the peak stress, at which point most bamboo fibers (with a low fiber volume fraction of 8%) fail, leading to a drop in stress. From this point on, the remaining coir fibers continue to bear some stress until the whole composite fails. When comparing the hybrid composites with the mono composites, the bamboo/PP composite fails in a brittle manner at high strength but low strain (<1%), and the coir/PP system shows a failure at low strength and somewhat higher strain; the E-modulus and strength of the hybrid composite are situated at intermediate values and there is, furthermore, some residual stress after the peak stress until higher strain values. This demonstrates a hybrid effect when combining strong bamboo fibers with high elongation coir fibers. Moreover, the failure strain of the bamboo fibers in the hybrid composite (~1.2%) is higher than in the mono-composite (~0.8%), suggesting that the presence of the coir fibers has a beneficial effect on the failure strain of the bamboo fibers. A possible explanation could be the stronger thermal contraction of the coir fibers during cooling after compression molding, leading to a mild compressive residual strain in the bamboo fibers. This effect has also been observed by several studies on hybrid composites [2,16–18].

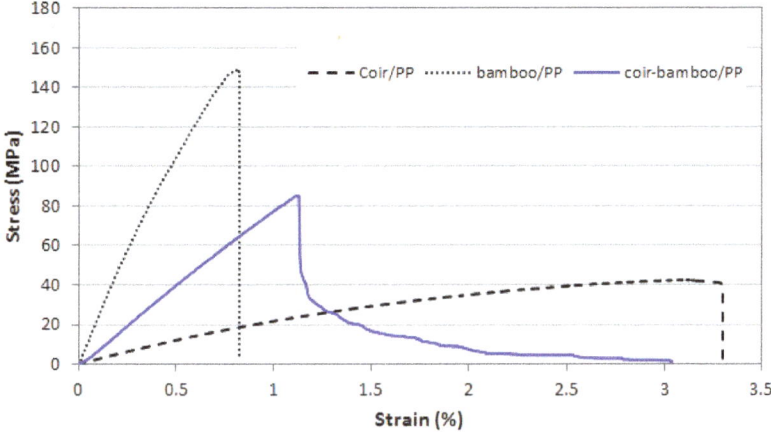

Figure 4. Typical tensile stress-strain curves of UD coir-bamboo/PP hybrid composites displayed together with stress-strain curves of UD mono coir/PP and UD mono bamboo/PP composites.

In (Figure 5b), the cross-section of the hybrid composite shows the distribution of the coir fibers and the bamboo fibers, which are still positioned in three different layers. This means that the hybrid effect in the composite is taking place at the meso level. The composite fracture shows many pulled out coir fibers and the presence of a few broken bamboo fibers. It is likely that the pull-out of the coir fibers delayed the failure of the composite as observed in its stress-strain curve.

The former is confirmed in (Table 2), which once more summarizes the tensile mechanical properties of the hybrid composites. The strain to failure of the hybrid composite is clearly higher than that of the mono bamboo/PP composite. The analysis of the tensile properties is further carried out by comparison with the theoretical values determined by the rule of mixtures.

Table 2. Tensile properties of coir-bamboo/PP hybrid composite, and of mono bamboo/PP and coir/PP composites.

Composites	V_{coir}(%)	V_{bamboo}(%)	E-Modulus (GPa)	Strength (MPa)	Strain at Failure (%)
Coir-bamboo/PP	30	8	7.3 ± 0.9	87.6 ± 4.4	2.2 ± 0.8
Bamboo/PP	0	45	21.7 ± 2.8	148.3 ± 10.5	0.9 ± 0.5
Coir/PP	44	0	2.4 ± 0.1	43.0 ± 0.8	3.3 ± 0.3

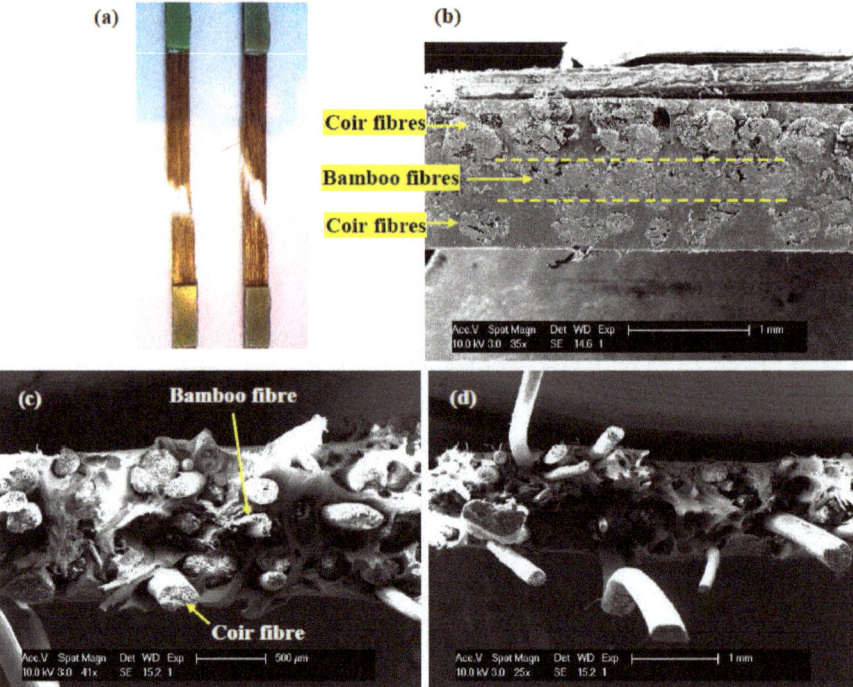

Figure 5. Fracture of the hybrid composite. (**a**) A typical sample fracture in tensile test; (**b**) a cross-section of the composite showing coir and bamboo fiber are distributed in three layers; (**c**) and (**d**) fracture surface of the composite in the tensile test.

3.2. Rule of Mixtures for the Hybrid Composite

3.2.1. Theoretical E-Modulus and Theoretical Strength of Mono-Composites

The theoretical E-modulus of coir/PP and bamboo/PP composites, E_{theo}, is calculated according to the rule of mixtures as shown in Equation (1):

$$E_{theo} = V_f E_f + (1 - V_f) E_m = V_f E_f + V_m E_m \tag{1}$$

where V_f and V_m are the volume fractions of fiber and matrix, respectively; E_f and E_m are the E-modulus of fiber and matrix, respectively.

In the coir and bamboo fiber composites with PP, the fiber failure strain (approximately 36% for coir and 1.9% for bamboo fibers) is lower than the matrix failure strain (higher than 300%). Thus, the strength of the fibers will determine the failure of the composites; hence the estimation of theoretical strength can be calculated as follows:

$$\sigma_{theo} = V_f \sigma_f^* + V_m \sigma_m' \tag{2}$$

where σ_f^* is the fiber strength, and σ_m' is the matrix stress at fiber failure strength.

3.2.2. Theoretical E-modulus and Theoretical Strength of Hybrid Composite

The theoretical E-modulus of the hybrid composite is calculated as:

$$E_{theo} = E_{coir}V_{coir} + E_{bamboo}V_{bamboo} + E_{PP}V_{PP} \tag{3}$$

As illustrated by Figure 6, the theoretical strength is estimated as follows:

(i) If V_{bamboo} is very low compared to V_{coir}: the strength of the composite is determined by coir fiber strength. In this case, the coir fibers can carry load after the failure of the bamboo fibers, then

$$\sigma_{theo} = \sigma^*_{coir}V_{coir} + \sigma'_{PP}V_{PP} \tag{4}$$

(ii) If V_{bamboo} is high. Then, the composite strength is dependent on bamboo fiber strength:

$$\sigma_{theo} = \sigma'_{coir}V_{coir} + \sigma^*_{bamboo}V_{bamboo} + \sigma'_{PP}V_{PP} \tag{5}$$

where σ'_{coir} and σ'_{PP} are the stress in the coir fiber and the stress in the PP, respectively, at the failure strain of the bamboo fiber (Figure 6).

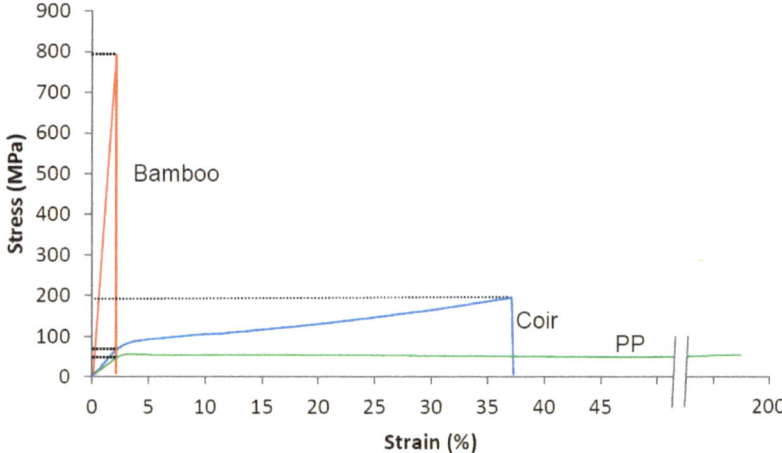

Figure 6. Mono-material properties used as input to calculate the properties following the rule of mixtures of the coir-bamboo/PP hybrid composite.

With the fiber volume fraction of coir fibers and of bamboo fibers at 30% and 8%, respectively, the theoretical strength of the composite calculated following Equation (4) is 111 MPa, which is lower than the value calculated following Equation (5) (120.8 MPa). The result shows the bamboo fiber load is high enough to determine the hybrid composite strength. Hence, the theoretical strength of the composite will be calculated according to Equation (5).

The theoretical E-modulus and strength of the hybrid composite is calculated following Equations (3) and (5) and shown in (Table 3). The efficiency factors (the experimental values normalized to the theoretical values) are also estimated. It can be seen that the strength efficiency factor is surprisingly high (0.73) compared to the values of the mono coir/PP (0.32) and bamboo/PP (0.38) systems. As discussed above, there likely exists a beneficial effect of the residual strain in bamboo fibers, leading to an important increase in failure stress and, hence, a higher contribution to the overall strength of the hybrid composite. Meanwhile, the premature failure of coir/PP and bamboo/PP (reflected with the relatively low strength efficiency factors) could be caused by low interfacial adhesion between the fibers and PP matrix.

Table 3. Theoretical E-modulus and strength of the hybrid composite estimated by the rule of mixtures, and the efficiency factors of E-modulus and strength.

Composite	Theoretical E-Modulus (GPa)	Efficiency Factor of E-Modulus	Theoretical Strength (MPa)	Efficiency Factor of Strength
Coir-bamboo/PP	6.1	1.22	120.8 [1]	0.73
Coir/PP	3.2	0.76	134.8	0.32
Bamboo/PP	23.5	0.92	393	0.38

[1] the theoretical strength of the composite is calculated according to Equation (5).

4. Conclusions

The tensile behavior of coir-bamboo fiber hybrid composites in PP was investigated, where the coir fiber and bamboo fiber were mixed at the meso level by layer-by-layer stacking of UD fiber prepregs. With a low bamboo fiber fraction, a hybrid effect with an increase of composite strain to failure was obtained, which can be attributed to the high strain to failure of the coir fibers; the bamboo fibers provided high stiffness and strength to the composites. The results show that a positive hybrid effect is obtained when a low bamboo fiber fraction is hybridized with a higher fraction of coir fibers. Different fiber mixing levels and variation of fiber loading can be considered to explore more synergetic properties for applications of the hybrid composites.

Author Contributions: Conceptualization: L.Q.N.T. and A.W.V.V.; methodology: L.Q.N.T. and A.W.V.V.; investigation: L.Q.N.T. and C.F.; writing: L.Q.N.T.; supervision: I.V.

Funding: This research received no external funding.

Conflicts of Interest: The authors declare no conflict of interest.

References

1. Swolfs, Y.; Verpoest, I.; Gorbatikh, L. Recent advances in fibre-hybrid composites: Materials selection, opportunities and applications. *Int. Mater. Rev.* **2019**, *64*, 181–215. [CrossRef]
2. Manders, P.W.; Bader, M. The strength of hybrid glass/carbon fibre composites. *J. Mater. Sci.* **1981**, *16*, 2246–2256. [CrossRef]
3. Zhang, J.; Chaisombat, K.; He, S.; Wang, C.H. Hybrid composite laminates reinforced with glass/carbon woven fabrics for lightweight load bearing structures. *Mater. Des.* **2012**, *36*, 75–80. [CrossRef]
4. Wisnom, M.R.; Czél, G.; Swolfs, Y.; Jalalvand, M.; Gorbatikh, L.; Verpoest, I. Hybrid effects in thin ply carbon/glass unidirectional laminates: Accurate experimental determination and prediction. *Compos. Part A Appl. Sci. Manuf.* **2016**, *88*, 131–139. [CrossRef]
5. Zhang, Y.; Li, Y.; Ma, H.; Yu, T. Tensile and interfacial properties of unidirectional flax/glass fiber reinforced hybrid composites. *Compos. Sci. Technol.* **2013**, *88*, 172–177. [CrossRef]
6. Swolfs, Y.; Crauwels, L.; Van Breda, E.; Gorbatikh, L.; Hine, P.; Ward, I.; Verpoest, I. Tensile behaviour of intralayer hybrid composites of carbon fibre and self-reinforced polypropylene. *Compos. Part A Appl. Sci. Manuf.* **2014**, *59*, 78–84. [CrossRef]
7. Pegoretti, A.; Fabbri, E.; Migliaresi, C.; Pilati, F. Intraply and interply hybrid composites based on E-glass and poly (vinyl alcohol) woven fabrics: Tensile and impact properties. *Polym. Int.* **2004**, *53*, 1290–1297. [CrossRef]
8. Hine, P.; Bonner, M.; Ward, I.M.; Swolfs, Y.; Verpoest, I.; Mierzwa, A. Hybrid carbon fibre/nylon 12 single polymer composites. *Compos. Part A Appl. Sci. Manuf.* **2014**, *65*, 19–26. [CrossRef]
9. Defoirdt, N.; Biswas, S.; De Vriese, L.; Van Acker, J.; Ahsan, Q.; Gorbatikh, L.; Van Vuure, A.; Verpoesta, L.; Trana, L.Q.N. Assessment of the tensile properties of coir, bamboo and jute fibre. *Compos. Part A Appl. Sci. Manuf.* **2010**, *41*, 588–595. [CrossRef]
10. Tran, L.Q.N.; Minh, T.N.; Fuentes, C.A.; Chi, T.T.; Van Vuure, A.W.; Verpoest, I. Investigation of microstructure and tensile properties of porous natural coir fibre for use in composite materials. *Indust. Crop. Prod.* **2015**, *65*, 437–445. [CrossRef]

11. Taketa, I.; Ustarroz, J.; Gorbatikh, L.; Lomov, S.V.; Verpoest, I. Interply hybrid composites with carbon fiber reinforced polypropylene and self-reinforced polypropylene. *Compos. Part A Appl. Sci. Manuf.* **2010**, *41*, 927–932. [CrossRef]
12. Taketa, I. Analysis of Failure Mechanisms and Hybrid Effects in Carbon Fibre Reinforced Thermoplastic Composites. Ph D. Thesis, Katholieke Universiteit Leuven, Leuven, Belgium, 2011.
13. Osorio, L.; Trujillo, E.; Van Vuure, A.W.; Verpoest, I. Morphological aspects and mechanical properties of single bamboo fibers and flexural characterization of bamboo/epoxy composites. *J. Reinf. Plast. Compos.* **2011**, *30*, 396–408. [CrossRef]
14. Osorio, L.; Trujillo, E.; Lens, F.; Ivens, J.; Verpoest, I.; Van Vuure, A.W. In-depth study of the microstructure of bamboo fibres and their relation to the mechanical properties. *J. Reinf. Plast. Compos.* **2018**, *37*, 1099–1113. [CrossRef]
15. Trujillo, E.; Moesen, M.; Osorio, L.; Van Vuure, A.W.; Ivens, J.; Verpoest, I. Bamboo fibres for reinforcement in composite materials: Strength Weibull analysis. *Compos. Part A Appl. Sci. Manuf.* **2014**, *61*, 115–125. [CrossRef]
16. Zweben, C. Tensile strength of hybrid composites. *J. Mater. Sci.* **1977**, *12*, 1325–1337. [CrossRef]
17. Bader, M.; Manders, P. The strength of hybrid glass/carbon fibre composites, part 1: Failure strain enhancement and failure mode. *J. Mater. Sci.* **1981**, *16*, 2233–2245.
18. Pitkethly, M.; Bader, M. Failure modes of hybrid composites consisting of carbon fibre bundles dispersed in a glass fibre epoxy resin matrix. *J. Phys. D Appl. Phys.* **1987**, *20*, 315. [CrossRef]

© 2019 by the authors. Licensee MDPI, Basel, Switzerland. This article is an open access article distributed under the terms and conditions of the Creative Commons Attribution (CC BY) license (http://creativecommons.org/licenses/by/4.0/).

Article

Bio-Polyamide 11 Hybrid Composites Reinforced with Basalt/Flax Interwoven Fibers: A Tough Green Composite for Semi-Structural Applications

Pietro Russo [1,*], **Giorgio Simeoli** [1], **Libera Vitiello** [2] **and Giovanni Filippone** [2,*]

1. Institute for Polymers, Composites and Biomaterials—National Council of Research, 80078 Pozzuoli, Naples, Italy; giorgio.simeoli@unina.it
2. Department of Chemical, Materials and Production Engineering, University of Naples Federico II, 80125 Naples, Italy; liberavitiello29@gmail.com
* Correspondence: pietro.russo@unina.it (P.R.); gfilippo@unina.it (G.F.)

Received: 11 March 2019; Accepted: 29 April 2019; Published: 6 May 2019

Abstract: Intraply hybrid green composites were prepared by film stacking and hot-pressing of bio-based polyamide 11 (PA11) sheets and commercial hybrid fabrics made by interweaving flax and basalt fibers (2/2 twill structure). Two matrices were considered, one of them containing a plasticizing agent. After preliminary thermal and rheological characterizations of the neat matrices, the laminates were studied in terms of flexural properties at low and high deformation rates, and the results were interpreted in the light of morphological analyses (scanning electron and optical microscopy). Despite the poor interfacial adhesion detected for all investigated composite samples, the latter exhibited a good combination of flexural strength, modulus, and impact resistance. Such well-balanced mechanical properties make the studied samples potential candidates for semi-structural applications, e.g., in the transportation sector.

Keywords: polyamide 11; interweaving flax-basalt fibers; hybrid composites; flexural properties

---

1. Introduction

In the last decades, a steady increase of interest has been devoted toward the design and development of hybrid composite systems given their outstanding perspectives of applications, even for advanced uses, deriving from the ability to combine advantages of the individual constituents, for example, stiffness and toughness, and the occurrence of synergisms, not yet well understood [1–4].

In line with the general trend to use reinforced plastics for both functional and structural applications, the hybridization approach, mainly obtained by embedding two or more different fibers within a polymer matrix, allows to tailor the properties of products to suit ever more specific requirements. In this regard, common configurations consist of different kinds of fibers distributed in different laminas or in the same one to form interply or intraply hybrids, respectively [5,6].

Gonzales et al. [7] focused on the low-velocity impact behavior of polymer-based interply hybrid laminates including woven carbon fabric, woven glass fabric, and unidirectional carbon tapes. Authors demonstrated that the stacking sequence of constituting layers can significantly affect results and failure mechanisms. In particular, the dissipation of impact energy is reduced when the woven fabrics are placed in the mid-plane of the studied composite structure, with a simultaneous increase of the residual properties. Ying et al. [8] studied the influence of hybridization on the impact properties of carbon-aramid/epoxy systems. Experimental tests highlighted that placing a high stiffness carbon fabric in correspondence of highly resistant regions permits to achieve enhanced properties of the reinforcement. Ferrante et al. [9] considered the effect of basalt fibers hybridization on the damage tolerance of carbon/epoxy laminates subjected to laser shock wave tests. The research indicated an optimal behavior for sandwich-like configurations, especially in the case of structures with basalt skins. Nisini et al. [10] analyzed ternary systems including carbon, basalt and flax fibers in an epoxy matrix obtained according to two different configurations. Samples were subjected to tensile, flexural, interlaminar shear strength and low-velocity impact tests. The inclusion of flax fibers showed a significant effect especially with regard to impact response of investigated materials.

Nowadays, the growing sensitivity towards environmental issues has increasingly moved the attention of researchers and industrials towards inherently recyclable materials and/or based on constituents coming from renewable sources. According to this consideration, the use of eco-sustainable thermoplastic resins and natural reinforcements is rapidly gaining a significant role even in industrial fields where the use of traditional carbon fiber- and glass fiber-reinforced thermosetting materials is widely established (aeronautics, naval, construction). Unfortunately, both natural fibers and low environmental impact plastics often suffer from poor performance, and their combination consequently results in "green composites" that do not meet the necessary requirements for many technologically relevant applications. The challenge is identifying new combinations of raw materials for the production of green composites whose performance is good enough to propose their use in suited applications. Among the various bio-based thermoplastic resins here we focus our attention on polyamide 11 (PA11), which is a semi-crystalline bio-polyamide produced using 11-aminoundecanoic acid derived from castor oil. Despite its relatively high costs, in the last decade, PA11 has gained a special industrial interest due to a good combination of mechanical properties and chemical resistance. In particular, PA11 exhibits good toughness compared to other bio-based thermoplastic resins, e.g., poly(lactic acid), which is often proposed as a matrix for bio-composites [11–15]. As far as the fibers are concerned, here we deal with a hybrid fabric made of basalt and flax interwoven fibers in 2/2 twill structure. This class of reinforcements has gained over time an extraordinary interest in research, as their use in polymeric matrices can offer different possibilities from the triggering of synergisms to mechanical properties not exhibited by composite materials similar but reinforced with each of the two combined fibers, separately [16–18]. The mechanical performance of composite laminates obtained by film stacking is investigated in both static and dynamic conditions. The results, appropriately supported by a morphological investigation of the induced damage, reveal a good combination of flexural properties and toughness, which suggest possible use in semi-structural applications, such as panels for the transportation field.

2. Materials and Methods

2.1. Materials

Two extrusion grade polyamide 11 (PA11) Besno Rilsan® from Arkema S.A. (Puteaux, France): a non-plasticised TL (density: 1.03 g/cm^3, MFI@235 °C/2.16 kg = 4.38 ± 1.25 g/10 min) and a plasticized P40 TL (density: 1.04 g/cm^3, MFI@235 °C/2.16 kg = 4.06 ± 0.61 g/10 min), were considered as matrices.

A hybrid fabric constituted by the interweaving of flax and basalt fibers in equal proportion and purchased at Lincore® (Bourguebus, France) with nominal areal weight 360 g/m^2 and an architecture twill 2/2 type was used as the reinforcement.

2.2. Laminates Preparation

Films with a thickness approximately equal to 80 µm were prepared using a Collin flat die extruder Teach-Line E20T equipped with a calender CR72T (Collin GmbH, Ebersberg, Germany). In detail, filming was conducted at a screw speed of 60 rpm, setting the temperature profile along the screw at 170–210–220–220–200 °C.

Composite laminates were obtained by the film stacking technique according to which PA11 films and hybrid fiber fabrics were alternately stacked and hot-pressed with a lab press Mod. P400E (Collin GmbH, Ebersberg, Germany) under pre-optimized conditions with a maximum temperature of 225 °C. The molding cycle is shown in Figure 1.

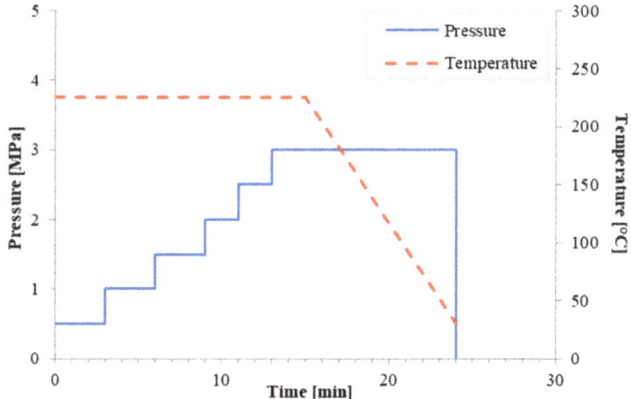

Figure 1. Processing conditions to prepare all PA11 based laminates.

Operating in this way, laminated samples were prepared by stacking four plies with plastic films. The final samples had a thickness approximately equal to 1.95 mm and volume fiber content of about 34%.

2.3. Calorimetric Tests

Differential scanning calorimetry (DSC) tests have been performed using a Q20 (TA Instruments, Milan, Italy) set-up on films of investigated matrices, at a rate of 10 °C/min and on the temperature range from 0 °C to 250 °C. The collected data permitted to assess the glass transition temperature of matrices and their degree of crystallinity according to the following Equation (1):

$$X_c = \frac{\Delta H_m}{\Delta H_{m0}} \tag{1}$$

where ΔH_m is the measured melting enthalpy while ΔH_{m0} is the melting enthalpy of the PA11 fully crystallized (189 J/g) [19].

2.4. Rheological Analysis

The flow behavior of the polymer matrices was investigated through rotational rheometry using a stress-controlled rotational rheometer (AR-G2 by TA Instruments) (Milan, Italy). Oscillatory shear experiments (frequency scans) were carried out to get the elastic (G′) and viscous (G″) moduli in the linear viscoelastic regime, whose limits were assessed through preliminary strain amplitude shear tests. The frequency scans were performed in air atmosphere from frequency $\omega = 10^2$ rad/s down to $\omega = 10^{-1}$ rad/s. The strain amplitude was $\gamma = 5\%$ for both PA11 TL and PA11 P40 TL resins. The testing temperatures were T = 220 °C for the PA11 TL and T = 205 °C for the PA11 P40 TL. This enabled to compare the rheological behavior of the two matrices at the same reduced temperature $\theta = T_{test} - T_{m-c} \approx 10$ °C (T_{m-c} being the temperature of melting peak closing). Note that lower temperatures were not explored because of the excessive viscosity of the selected matrices.

2.5. Morphological Observations

Morphological analysis was conducted on cryo-fractured surfaces of composite samples to highlight any interfacial feature useful to support mechanical results. In this regard, observations were captured with a field emission scanning electron microscope (mod. FEI QUANTA 200 F) (Zurich, Switzerland) operating in high vacuum conditions at the voltage of 20 kV. Analyzed surfaces were previously coated with a thin layer of a gold-palladium alloy.

2.6. Static-Mechanical Tests

Tensile and flexural tests were conducted with a Tensometer 2020 (Alpha Technologies, Cinisello Balsamo, Milan, Italy).

In particular, tensile measurements on the neat polymer matrices were performed on dog-bone shaped specimens, obtained by compression molding, according to the ASTM D638-14. Five specimens were tested at room temperature, with a displacement rate of 5 mm/min and using a load cell of 10 kN.

Flexural tests were carried out on both neat matrices and laminates. The measurements were performed loading each specimen up to 5% of strain, according to the ASTM D790-03, by using a load-cell of 500 N. The reported results represent average values computed from five independent measurements per each sample.

2.7. Charpy-Like Tests

High-velocity flexural properties of the laminates were estimated by means of an instrumented Charpy impact testing machine CEAST 9500 (ITW Test and Measurements, Pianezza, Turin, Italy). Three-point bending tests were performed on five specimens with a length approximately equal to 100 mm and using a span width of 62 mm, at a load application speed of 3.8 m/s. Results are reported in terms of stress-deformation curves.

3. Results

The main calorimetric properties collected through DSC analysis are summarized in Table 1. The plasticizing agent present in the sample PA11 P40 TL causes the reduction of the glass transition temperature, which passes from T_g = 50.5 °C to T_g = 37.7 °C. Besides this expected reduction of T_g, the plasticizer also affects the melting peak, which results narrowed and shifted to lower temperatures, with obvious advantages in terms of processability. Furthermore, it is interesting to notice that the melting enthalpy of the two samples is essentially the same and, according to the Equation (1), it corresponds to a degree of crystallinity approximately equal to $\chi \approx 27\%$.

Table 1. Main calorimetric properties of the polymer matrices.

Parameter	PA11 TL	PA11 P40 TL
Glass transition temperature, T_g in °C	50.5	37.7
Melting temperature, T_m (onset/peak/peak closing) in °C	182.6/193.8/210.2	164.7/182.5/195.5
Melting enthalpy, ΔH_m in J/g	50.6	51.3
Degree of crystallinity, χ in %	26.7	27.1

The results of rheological analyses are shown in Figure 2a,b, where G' and G" are shown as a function of frequency together with the complex viscosity, $\eta^* = \sqrt{G'^2 + G''^2}/\omega$. When compared at the same reduced temperature θ, the samples are almost indistinguishable and share the same phenomenological behavior, characterized by moduli comparable between them in the whole range of investigated frequency. G" is slightly higher than G' at low frequency, while the moduli cross each other at $\omega \approx 10$ rad/s and G' exceeds G" for higher frequency. Both moduli apparently approach a plateau value at low frequency. Actually, such behavior is a consequence of polymer degradation during time, which causes a growth of the viscoelastic quantities while testing from high to low frequencies. This can be seen in the insets of Figure 2a,b, where the complex viscosity at $\omega \approx 10$ rad/s is reported as a function of time. In this regard, two aspects can be highlighted: (i) the viscosity of the selected polymers, which can be assumed equivalent to η^* according to the Cox–Merz rule, is very high, being of order of ~10^4 Pa s in the range of shear rate typically experienced during the film stacking step for the preparation of the laminates (i.e., $\dot{\gamma} \sim 5$ 1/s, see [20]), (ii) the viscosity grows quite rapidly over time. This rheological information is precious when considering the efficacy of the film stacking step and the mechanical performance of the laminates. First of all, the film stacking process was carried out at T = 190 °C, which is a lower temperature than that of rheological tests. It is hence reasonable to expect that the viscosity of both matrices during laminate preparation was even higher than that measured via rheological tests. The effect of such a high viscosity of the matrices is a delay in the permeation times of the polymer in the fabric, which inversely depends on the viscosity as predicted by the Darcy's law for the flow in porous mediums. Slow permeation times can have detrimental effects on the level of compaction of the laminates. On the other hand, the solution of prolonging the duration of the hot pressing process is not feasible in this case because of the growth of the viscosity over time (see insets of Figure 2a,b).

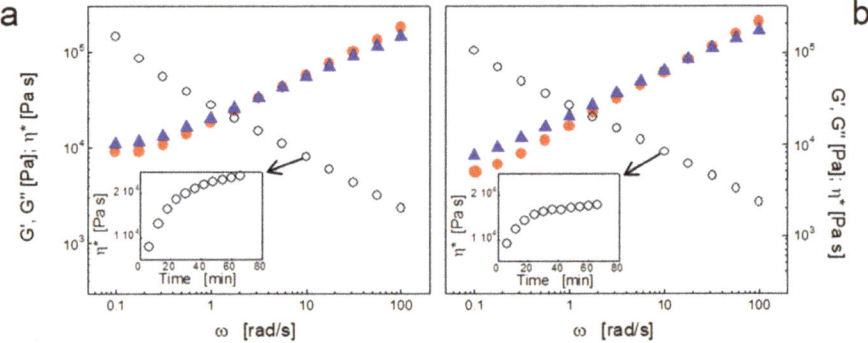

Figure 2. Frequency dependence of G′ (red circles), G″ (blue triangles), and η * (empty circles) for PA11 TL (**a**) and PA11 P40 TL (**b**) at θ ≈ 10 °C. The time evolution of G′ (red circles) and G″ (blue triangles) at ω = 10 rad/s is shown in the insets.

Figure 3 shows the tensile stress-strain curves of considered polyamides. Clearly, both matrices exhibit a ductile behavior with the non-plasticized polymer showing, as expected, higher tensile modulus and strength, as well as lower deformation at break than the plasticized one.

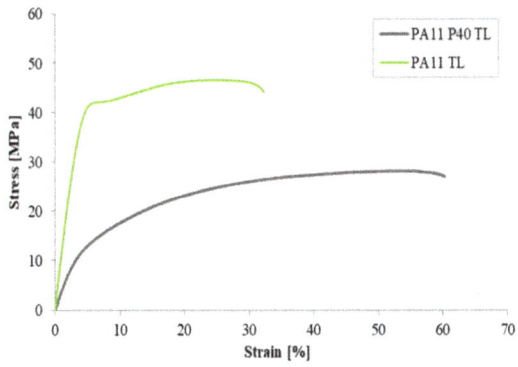

Figure 3. Representative tensile stress-strain curves of PA11 matrices.

The tensile parameters obtained by processing these curves and summarized in Table 2 are in line with data already available in the literature [21].

Table 2. Tensile parameters.

Parameter	PA11 TL	PA11 P40 TL
Young modulus, in GPa	1.01 ± 0.02	0.39 ± 0.02
Yield stress (0.2% offset), in MPa	38.8 ± 3.6	15.5 ± 0.4
Yield strain (0.2% offset), in %	4.9 ± 0.5	7.5 ± 0.5
Tensile strength, in MPa	44.9 ± 0.5	29.4 ± 1.1
Stress at break, in MPa	43.3 ± 2.4	28.8 ± 1.5
Strain at break, in %	26.9 ± 6.6	66.0 ± 5.8
Toughness, in MPa	10.7 ± 3.4	15.9 ± 1.6
Resilience, in MPa	1.56 ± 0.01	0.74 ± 0.08

Similarly, Figure 4 refers to the flexural response of PA11 TL and PA11 P40 TL. In this case, both resins show that the stress, at least on the deformation range foreseen by the reference standard (up to 5%), is continuously increasing. This behavior can be explained on the basis of the ductility of the studied materials and, therefore, of their ability to support the load without yielding until 5% of deformation is reached.

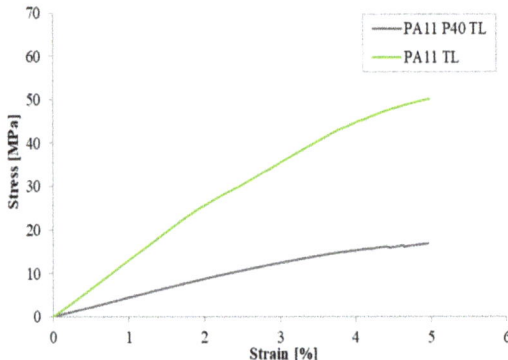

Figure 4. Representative flexural stress-strain curves of PA11 matrices.

Table 3, showing the average values and standard deviations of both the elastic modulus and the flexural strength for the two polyamide matrices, confirms that the presence of plasticizers implies, even in the case of flexural loads, a significant reduction in performance. In particular, the flexural stiffness and strength of the plasticized sample are about three times lower respect to the not plasticized one.

Table 3. Flexural parameters of PA11 matrices.

Parameter	PA11 TL	PA11 P40 TL
Flexural modulus, in GPa	1.31 ± 0.11	0.45 ± 0.06
Flexural offset yield strength (0.2% offset), in MPa	30.2 ± 2.4	11.8 ± 1.5
Yield strain (0.2% offset), in %	2.55 ± 0.13	2.72 ± 0.11
Flexural strength, in MPa	>50.3	>17.3

Once the polymer matrices have been characterized, the attention was moved on the laminates. First of all, SEM analyses were performed to investigate the interactions between the fibers and the matrices. The micrographs of the cryo-fractured surfaces of the composite systems are shown in Figure 5. The pictures clearly highlight a poor interfacial adhesion, which suggests the propensity of both the investigated composite systems to undergo dissipative phenomena, such as delamination, when subjected to external loads.

Figure 5. SEM micrographs of (**a**) PA11 TL and (**b**) PA11 P40 TL based composites.

The composites were subjected to flexural tests, and the results are shown in Figure 6 in terms of representative stress-strain curves. The numerical results are resumed in the Table 4.

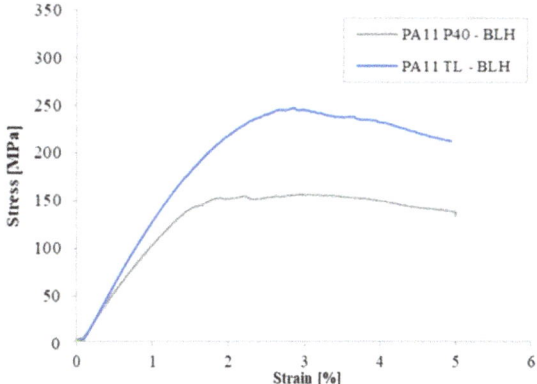

Figure 6. Representative flexural stress-strain curves of PA11 hybrid composites.

Table 4. Flexural parameters of PA11 based composites.

	PA11 TL-BLH	**PA11 P40 TL-BLH**
Flexural modulus, in GPa	9.1 ± 0.4	7.5 ± 0.2
Flexural offset yield strength (0.2% offset), in MPa	173 ± 8.2	90 ± 7.4
Yield strain (0.2% offset), in %	1.46 ± 0.24	0.82 ± 0.16
Flexural strength, in MPa	250.1 ± 9.4	159.3 ± 6.1

First of all, none of the samples broke within 5% of strain, that is the upper limit of strain envisaged by ASTM D790-03. This proves the high toughness of the investigated composites. Regarding the comparison between the two samples, the stress-strain curves share the same qualitative behavior, reaching a maximum before a decrease of the sustained stress. Looking at the numerical values in Table 4, it is interesting to observe that the flexural modulus and strength are higher than what reported in the literature for many green composites, and they are in line with the benchmark of glass fiber-reinforced ordinary laminates [22].

The high toughness of the samples was further proved by means of high-velocity flexural tests. Specifically, Charpy-like tests were performed considering an anomalous configuration in which the specimens were struck on the width side to simulate a high speed three-point flexural test. The schematic of the experimental set up is shown in Figure 7a. The results are shown in Figure 7b for representative samples in terms of stress versus strain, where the latter was computed from the deflection data according to the following equation:

$$\varepsilon_f = \frac{6Dd}{L^2} \qquad (2)$$

where D is the displacement, L is the span length, and d is the sample thickness.

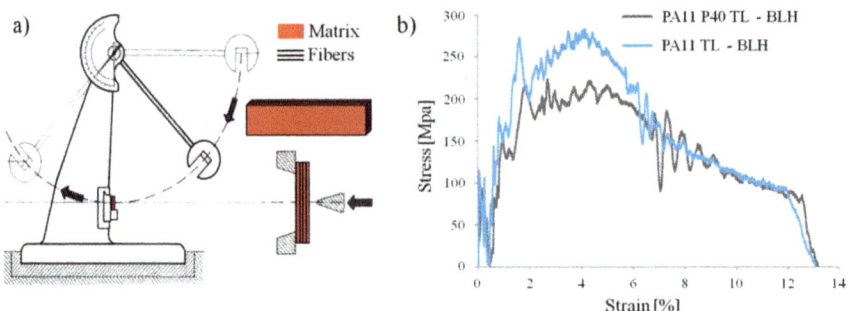

Figure 7. (**a**) Experimental set-up for the Charpy-like tests. (**b**) High speed-flexural stress-strain curves of the investigated PA11 composite laminates.

First of all, the results indicate that all specimens did not break during the high-speed flexural test. This further proves the high toughness of the investigated samples, which bend without breaking eventually slipping away. An estimate of the maximum deformation experienced by the samples can be computed through simple geometrical considerations, which lead to a limit deflection of about 39 mm. When properly converted with Equation (2), this value corresponds to about $\varepsilon_f \approx 12\%$ of percentage strain. From a qualitative point of view, the two samples share the same behavior: the stress sharply increases until a maximum at about 4% of strain, and then it gradually decreases. No consideration can be made about the comparison between the rigidity of investigated composite systems, given the usual noise of this kind of curves. However, the performance of the composite based on PA11 P40 TL is only slightly lower than those of the sample based on the PA11 TL in spite of the big differences in terms of the flexural properties of the two matrices.

In order to elucidate the mechanism on the basis of the appreciable flexural properties of the composites based on the plasticized matrix, morphological observations were collected by an optical microscope at the end of the Charpy-like tests. In particular, pictures were taken on the damaged area of impacted specimens, and precisely on the strike face, subjected to compression load, on the corresponding rear side, experiencing tension, and along the thickness direction (Figure 8).

The pictures show that the not plasticized PA11 TL matrix shows clear evidence of damage on the front and rear surfaces of the sample (Figure 8a, picture A and B, respectively), subjected to compression and tension stresses during the impact, respectively. In contrast, the composite based on the tougher PA11 P40 matrix resists without severe damages on the surface polymeric skins (Figure 8b, picture A and B). As a result of the higher flexibility of the plasticized matrix, the breaking mechanisms of fibers are relieved, as well as the crack propagation in the matrix.

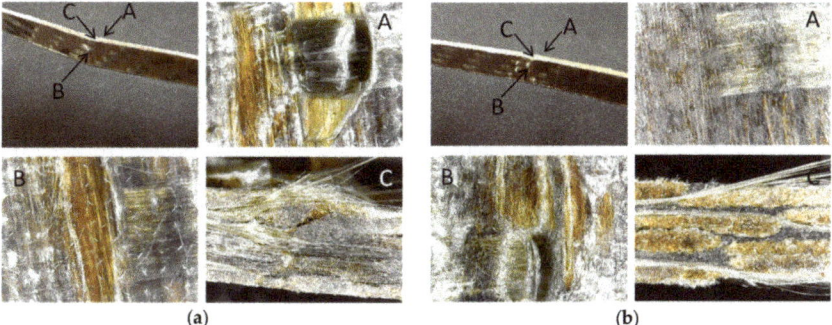

Figure 8. Optical micrographs showing the composite samples based on the not plasticizes PA11 TL (**a**) and plasticized PA11 P40 (**b**) after the Charpy-like tests. For both figures, the sub figures A, B and C represent the top view, the bottom view and the side view, respectively.

4. Conclusions

Hybrid composite laminates constituted by embedding an interweaved flax and basalt fibers fabric in two polyamide 11 resins, pure and plasticized, were prepared by film stacking and hot-pressing techniques under processing conditions preliminarily optimized by thermal and rheological analyses of the matrices. Specimens of appropriate size, cut from both pure matrix sheets and composite laminates, were subjected to mechanical tests. Specifically, pure resin specimens were studied by quasi-static tensile and flexural tests, while composite specimens were subjected to flexural measurements carried out at both low and high strain rate.

The plasticized matrix exhibited lower stiffness compared to its not plasticized counterpart, but its toughness was more than 50% higher. SEM investigations highlighted a poor polymer-fibers interfacial adhesion in both composites, with detrimental effects in terms of stress transfer ability, but with possible benefits in terms of dissipative phenomena under large deformations. The latter were explored through flexural tests performed at both low and high velocity. When tested under quasi-static conditions, the composites exhibited flexural stiffness and strength higher than those reported in the literature for green composites, and comparable with those of glass-fiber reinforced laminates. Good flexural properties were maintained even under high-velocity conditions, especially for the sample with a plasticized matrix. Overall, the present study demonstrates the suitability of the investigated green composites based on hybrid reinforcement for the manufacturing of items that must meet at least semi-structural requirements, e.g., for panels in the transportation field.

Author Contributions: Conceptualization, P.R. and G.F.; Methodology, P.R.; Formal analysis, P.R. and G.F.; Investigation, L.V. and G.S.; Writing—original draft preparation, P.R. and G.F.; Writing—review and editing, P.R. and G.F.; Supervision, P.R. and G.F.

Funding: This research received no external funding.

Conflicts of Interest: The authors declare no conflict of interest.

References

1. Maries, I.; Kuruvilla, J.; Sabu, T. Mechanical performance of short banana/sisal hybrid fiber reinforced polyester composites. *J. Reinf. Plast. Comp.* **2010**, *29*, 12–29.
2. Manikandan, P.; Chai, G.B. A similitude approach towards the understanding of the low velocity impact characteristics of bi-layered hybrid composite structures. *Comp. Struct.* **2015**, *131*, 183–192. [CrossRef]
3. Petrucci, R.; Santulli, C.; Puglia, D.; Nisini, E.; Sarasini, F.; Tirillò, J.; Torre, L.; Minak, G.; Kenny, J.M. Impact and post-impact damage characterisation of hybrid composite laminates based on basalt fibres in combination with flax, hemp and glass fibres manufactured by vacuum infusion. *Comp. Part B* **2015**, *69*, 507–515. [CrossRef]

4. Swolfs, Y.; Gorbatikh, L.; Verpoest, I. Fibre hybridization in polymer composites: A review. *Comp. Part A* **2014**, *67*, 181–200. [CrossRef]
5. Attia, M.A.; Abd El-Baky, M.A.; Alshorbagy, A.E. Mechanical performance of intra-ply and inter-intraply hybrid composites based on e-glass and polypropylene unidirectional fibres. *J. Comp. Mater.* **2017**, *51*, 381–394. [CrossRef]
6. Audibert, C.; Andreani, A.S.; Laine, E.; Grandidier, J.-C. Mechanical characterization and damage mechanism of a new flax-kevlar hybrid/epoxy composite. *Comp. Struct.* **2018**, *195*, 126–135. [CrossRef]
7. Gonzalez, E.V.; Maimi, P.; Sainz de Aja, J.R.; Cruz, P.; Camanho, P.P. Effects of interply hybridization on the damage resistance and tolerance of composite laminates. *Comp. Struct.* **2014**, *108*, 319–331. [CrossRef]
8. Ying, S.; Mengyun, T.; Zhijun, R.; Baohui, S.; Li, C. An experimental investigation on the low-velocity impact response of carbon-aramid/epoxy hybrid composite laminates. *J. Reinf. Plast. Comp.* **2017**, *36*, 422–434. [CrossRef]
9. Ferrante, L.; Tirillò, J.; Sarasini, F.; Touchard, F.; Ecault, R.; Vidal Urriza, M.A.; Chocinski-Arnault, L.; Mellier, D. Behaviour of woven hybrid basalt-carbon/epoxy composites subjected to laser shock wave testing: Preliminary results. *Comp. Part B Eng.* **2015**, *78*, 162–173. [CrossRef]
10. Nisini, E.; Santulli, C.; Liverani, A. Mechanical and impact characterization of hybrid composite laminates with carbon, basalt and flax fibres. *Comp. Part B Eng.* **2017**, *127*, 92–99. [CrossRef]
11. Lebaupin, Y.; Chauvin, M.; Truong Hoang, T.-Q.; Touchard, F.; Beigbeder, A. Influence of constituents and process parameters on mechanical properties of flax fibre-reinforced polyamide 11 composite. *J. Therm. Comp. Mater.* **2017**, *30*, 1503–1521. [CrossRef]
12. Oliver-Ortega, H.; Llop, M.F.; Espinach, F.X.; Tarres, Q.; Ardanuy, M.; Mutje, P. Study of the flexural modulus of lignocellulosic fibers reinforced bio-based polyamide 11 green composites. *Comp. Part B* **2018**, *152*, 126–132. [CrossRef]
13. Oliver-Ortega, H.; Mendez, J.A.; Reixach, R.; Espinach, F.X.; Ardanuy, M.; Mutjé, P. Towards more sustainable material formulations: A comparative assessment of PA11-SGW flexural performance versus oil-based composites. *Polymers* **2018**, *10*, 440. [CrossRef] [PubMed]
14. Armioun, S.; Panthapulakkal, S.; Scheel, J.; Tjong, J.; Sain, M. Biopolyamide hybrid composites for high performance applications. *J. Appl. Polym. Comp.* **2016**, *133*, 43595. [CrossRef]
15. Haddou, G.; Dandurand, J.; Dantras, E.; Maiduc, H.; Thai, H.; Vu Giang, N.; Huu Trung, T.; Ponteins, P.; Lacabanne, C. Mechanical properties of continuous bamboo fiber-reinforced biobased polyamide 11 composites. *J. Appl. Polym. Comp.* **2019**, *136*, 47623. [CrossRef]
16. Fragassa, C.; Pavlovic, A.; Santulli, C. Mechanical and impact characterisation of flax and basalt fibre vinylester composites and their hybrids. *Compos. Part B Eng.* **2018**, *137*, 247–259. [CrossRef]
17. Fiore, V.; Calabrese, L.; Di Bella, G.; Scalici, T.; Galtieri, G.; Valenza, A.; Proverbio, E. Effects of ageing in salt spray conditions of flax and flax/basalt reinforced composites: Wettability and dynamic-mechanical properties. *Comp. Part B Eng.* **2016**, *93*, 35–42. [CrossRef]
18. Zivkovic, I.; Fragassa, C.; Pavlovic, A.; Brugo, T. Influence of moisture absorption on the impact properties of flax, basalt and hybrid flax/basalt fibre reinforced green composites. *Comp. Part B Eng.* **2017**, *111*, 148–164. [CrossRef]
19. Zhang, Q.; Mo, Z.; Liu, S.; Zhang, H. Influence of annealing on structure of Nylon 11. *Macromolecules* **2000**, *33*, 5999–6005. [CrossRef]
20. Agarwal, G.; Reyes, G.; Mallick, P. Study of compressibility and resin flow in the development of thermoplastic matrix composite laminates by film stacking technique. In *ASC Series on Advances in Composite Materials*; Loos, A., Hyer, M.W., Eds.; DESTech Publications: Lancaster, PA, USA, 2013; Volume 6, pp. 215–227.
21. Rilsan Polyamide 11 Brochure. Available online: https://www.extrememateriels-arkema.com/export/sites/technicalpolymers/.content/medias/downloads/brochures/rilsan-brochures/Rilsan-Polyamide-11-Brochure-optimized.pdf (accessed on 1 March 2019).
22. Pickering, K.L.; Efendy, M.A.; Le, T.M. A review of recent developments in natural fibre composites and their mechanical performance. *Comp. Part A Appl. Sci. Manuf.* **2016**, *83*, 98–112. [CrossRef]

© 2019 by the authors. Licensee MDPI, Basel, Switzerland. This article is an open access article distributed under the terms and conditions of the Creative Commons Attribution (CC BY) license (http://creativecommons.org/licenses/by/4.0/).

Article

Quasi-Static and Low-Velocity Impact Behavior of Intraply Hybrid Flax/Basalt Composites

Fabrizio Sarasini [1,*], Jacopo Tirillò [1], Luca Ferrante [1], Claudia Sergi [1], Pietro Russo [2], Giorgio Simeoli [2], Francesca Cimino [2], Maria Rosaria Ricciardi [3] and Vincenza Antonucci [3]

1. Department of Chemical Engineering Materials Environment, Sapienza-Università di Roma and UdR INSTM, Via Eudossiana 18, 00184 Roma, Italy; jacopo.tirillo@uniroma1.it (J.T.); luca.ferrante@uniroma1.it (L.F.); claudia.sergi@uniroma1.it (C.S.)
2. Institute for Polymers, Composites and Biomaterials, National Research of Council, Via Campi Flegrei 34, 80078 Pozzuoli (NA), Italy; pietro.russo@unina.it (P.R.); giorgio.simeoli@unina.it (G.S.); francesca.cimino@ipcb.cnr.it (F.C.)
3. Institute for Polymer, Composites and Biomaterials, National Research Council, Piazzale Enrico Fermi, 1, 80055 Portici (NA), Italy; mariarosaria.ricciardi@unina.it (M.R.R.); vinanton@unina.it (V.A.)
* Correspondence: fabrizio.sarasini@uniroma1.it; Tel.: +39-0644585408

Received: 20 February 2019; Accepted: 19 March 2019; Published: 22 March 2019

Abstract: In an attempt to increase the low-velocity impact response of natural fiber composites, a new hybrid intraply woven fabric based on flax and basalt fibers has been used to manufacture laminates with both thermoplastic and thermoset matrices. The matrix type (epoxy or polypropylene (PP) with or without a maleated coupling agent) significantly affected the absorbed energy and the damage mechanisms. The absorbed energy at perforation for PP-based composites was 90% and 50% higher than that of epoxy and compatibilized PP composites, respectively. The hybrid fiber architecture counteracted the influence of low transverse strength of flax fibers on impact response, irrespective of the matrix type. In thermoplastic laminates, the matrix plasticization delayed the onset of major damage during impact and allowed a better balance of quasi-static properties, energy absorption, peak force, and perforation energy compared to epoxy-based composites.

Keywords: flax fibers; basalt fibers; intraply flax/basalt hybrid; low-velocity impact; mechanical properties

1. Introduction

The need to increase the mechanical performance of natural fiber composites to meet the requirements of at least semi-structural applications has triggered a resurgent interest in hybrid composites [1,2]. Indeed, fiber hybridization offers a comprehensive set of possibilities leading to synergetic effects or to properties not exhibited by the single constituents [3]. This approach has been successfully exploited in the field of low-velocity impact resistance of composite laminates [4], with the first studies aimed at increasing the damage tolerance of carbon fiber composites by adding more ductile fibers, mainly glass [5–7] and aramid [8–12]. Recently new fiber combinations have been investigated, mainly based on natural and synthetic fibers. Not only glass fibers [13,14] but also carbon fibers [15,16] have been successfully hybridized with natural fibers to enhance poor mechanical properties and moisture resistance of natural fiber composites.

Among the different combinations available, the use of flax fibers with natural fibers of mineral origin, such as basalt fibers [17], has been widely investigated in literature with promising results. Fragassa et al. [18] addressed the impact behavior of hybrid laminates made of basalt and flax fibers in a vinylester matrix. These hybrid laminates were manufactured using a sandwich-like configuration with softer flax fibers in the core layers and the stronger basalt ones in the skins, in accordance with a stacking sequence typically used in impact resistant hybrid laminates [19].

Aside from intermediate tensile and flexural properties between those of basalt and flax fiber laminates, hybrid laminates showed a higher penetration energy compared to pure flax fiber composites. Recently Papa et al. [20] investigated the impact response of hybrid basalt/flax epoxy composites with an intercalated stacking sequence ($[B,F]_{8s}$). Hybridization led to better impact performance compared to pure basalt and flax composites in terms of peak force and penetration energy along with a much lower delaminated area. This behavior was due, according to the authors, to the energy absorption ability of flax layers through a non-elastic mode and to the deflection of the impact damage progression. From the above-mentioned studies, it is evident that material dispersion can be considered as one of the most important parameters in hybridization under low-velocity impact loading [1].

In particular, it is important to differentiate the intraply hybridization, where yarns (tows) of two different fibers are mixed in the same ply, from the interply one, where plies of two homogeneous reinforcements are stacked. It is expected that intraply hybrid composites exhibit better resistance to crack propagation during an impact event, even though results in the literature are contrasting [21,22].

Compared to interply hybrid composites, which have been widely investigated in literature, intraply hybrid composites have received limited attention. Zhang et al. [23] studied the effects of different hybrid structures based on interlayer and intralayer warp-knitted fabrics with carbon and glass fibers under low-velocity impact conditions. Compared to the interlayer configuration, the intralayer hybrid showed a higher peak load and a smaller damage area at the same hybrid ratio and level of impact energy, thus, suggesting that a better impact resistance can be obtained by intralayer hybrid structures. Bandaru et al. [24] addressed the low-velocity impact response of 3D angle-interlock polypropylene composites reinforced with Kevlar and basalt fibers, and the intraply configuration was found to absorb more energy (7.67–48.49%) than the pure 3D Kevlar and basalt composites. Wang et al. [22] manufactured three-dimensional interply and intraply basalt–Kevlar/epoxy hybrid composites. The low-velocity impact test results demonstrated higher ductility indices, lower peak load, and higher specific energy absorption in both warp and weft directions of the interply hybrid composite compared to those of the intraply hybrid composite. In addition to the material dispersion, the impact behavior of a composite material is also significantly affected by matrix type and its toughness. Thermoplastic matrices are usually preferred over thermoset ones when an improvement in impact resistance and damage tolerance is required [25,26].

In this work, an intraply technology is presented, and this could represent an interesting solution for introducing flax fibers in at least semi-structural components. This work aims to evaluate the effect of two different matrices, namely a thermoset (epoxy) and a thermoplastic (polypropylene), on the mechanical properties of a new hybrid composite based on a commercially available hybrid woven fabric with basalt and flax fibers. The hybrid composites have been subjected to quasi-static flexural tests along with low-velocity impacts to investigate their mechanical behavior and correlate the resulting failure modes. To the best of authors' knowledge, no similar works are available combining, in a single study, this specific intraply hybridization with different polymer matrices.

2. Materials and Methods

2.1. Materials

LINCORE® HF T2 360 (Figure 1), provided by Depestele (Bourguebus, France), was a balanced Twill 2/2 woven fabric with an areal density of 360 g·m^{-2} (50 wt% flax/50 wt% basalt; threads per cm in warp/weft = 3.6). Twill fabrics are common in the composite industry as they show longer thread flotation compared to plain woven fabrics and a lower level of fiber crimp, potentially leading to better mechanical properties and higher fiber packing density along with a better ability to conform to complex contours.

Figure 1. Close-up view of the dry flax/basalt hybrid fabric (dark yarn = basalt; bright yarn = flax).

The thermoset matrix was based on a two-component commercial epoxy system PRIME™ 20LV (100:26 resin/hardener weight ratio) supplied by Gurit (Newport, UK). The thermoplastic matrix was a polypropylene (PP, Bormod HF955MO) with a MFI@230 °C, 2.16 kg equal to 20 g/10 min. To increase the interfacial adhesion with both reinforcements, a PP-g-MA (maleic anhydride grafted polypropylene) coupling agent (2 wt%) from Chemtura (Philadelphia, USA) was used, namely Polybond 3000 (MFI@190 °C, 2.16 kg: 405 g/10 min; maleic anhydride content of 1.2 wt%). Polyolefins, such as polyethylene and polypropylene are known to exhibit a poor adhesion with natural fibers due to their hydrophobic long aliphatic primary chain without any polar groups. PP-g-MA coupling agent was chosen to react with the hydroxyl groups on the surface of both basalt and flax fibers [27,28].

The commercial polypropylene (PP) was modified with a coupling agent (PPc) by a co-rotating twin screw extruder Collin Teach-Line ZK25T with the following temperature profile: 180–190–205–195–185 °C from the hopper to the die, and with a screw speed of 60 rpm. Films of neat or compatibilized polypropylene with 2 wt% of coupling agent, with a thickness equal to 35–40 μm, were obtained by using a film blowing extrusion line model Teach-Line E 20 T from Collin GmbH (Maitenbeth, Germany) equipped with a calender CR72T (Maitenbeth, Germany). The processing was performed in accordance with the following temperature profile along the screw: 180–190–200–190–185 °C and a screw speed of 55 rpm.

2.2. Composite Materials Manufacturing

Thermoset-based laminates have been manufactured by vacuum infusion process by stacking 4 layers (0/90) of the hybrid fabric that were cured for 16 h at 50 °C as per manufacturer's specifications. Thermoplastic laminates were manufactured by alternating layers of polypropylene films (with or without coupling agent) and 4 hybrid fabrics by the film-stacking technique using a compression molding machine (model P400E, Collin GmbH, Maitenbeth, Germany) in accordance with the pre-optimized molding cycle shown in Figure 2. In Table 1 the characteristics of the as-manufactured composite materials are summarized.

Table 1. Characteristics of the as-manufactured hybrid composites.

Material ID	Matrix Type	Total Fiber Volume Fraction	Thickness (mm)
H_Ep	Thermoset-Epoxy	0.38 ± 0.02	2.0 ± 0.1
H_PP	Thermoplastic-PP	0.36 ± 0.02	2.1 ± 0.1
H_PPc	Thermoplastic-PPc	0.35 ± 0.03	2.2 ± 0.1

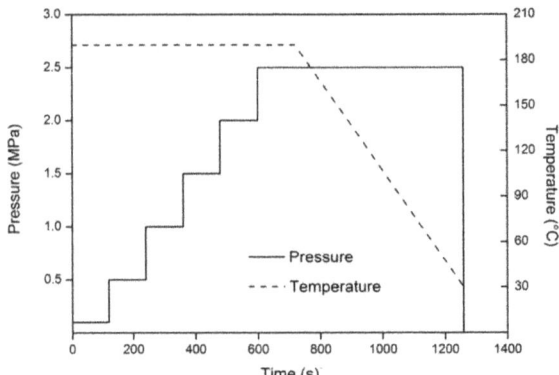

Figure 2. Processing conditions used to manufacture polypropylene (PP)- and polypropylene modified with a coupling agent (PPc)- based composites.

2.3. Mechanical Characterization of Composites

Specimens with dimensions 125 × 20 × t ($l \times w \times t$) were subjected to three-point bending tests in accordance with ASTM D790 (West Conshohocken, PA, USA) with a support span-to-thickness ratio of 32:1 and a cross-head speed of 5 mm/min and 2.5 mm/min for thermoplastic and thermoset-based composites, respectively. Tests were carried out on a Zwick/Roell Z010 universal testing machine (Ulm, Germany) equipped with a 10 kN load cell. The flexural strain was measured by a sensor arm for flexure test that was connected with an automatic extensometer. Specimens were tested in two orientations: with flax fibers along the longitudinal direction (warp fiber direction, F_L) and with basalt fibers along the longitudinal direction (weft fiber direction, B_L). Five specimens were tested for each configuration and matrix type.

Test coupons measuring 100 × 100 mm were impacted at room temperature at target kinetic impact energies ranging from 5 J to 30 J. An instrumented drop-weight impact testing machine (Instron/CEAST 9340, Pianezza, Italy) was used to this purpose equipped with a hemispherical tip (diameter of 12.7 mm). Three specimens for each matrix type and energy level were impacted out-of-plane with a constant mass of 8.055 kg while being clamped between two steel plates with a circular unsupported area of 40 mm (diameter). The force–time curves were recorded during each test by the DAS64K acquisition system.

2.4. Morphological and Damage Investigation

The fracture surfaces of specimens which failed in bending were investigated by scanning electron microscopy (FE-SEM MIRA3 by TESCAN, Brno, Czech Republic). All specimens were sputter coated with gold prior to FE-SEM observations.

The dent depth of each impacted coupon was measured using a laser profilometer (Taylor–Hobson Talyscan 150) with a scanning speed of 8500 μm/s. The scanned images were processed with the analysis software TalyMap 3D.

3. Results and Discussion

3.1. Quasi-Static Flexural Behaviour

The flexural properties are summarized in Table 2, while Figure 3 shows representative stress–strain curves obtained at room temperature.

Table 2. Summary of flexural properties of flax/basalt hybrid laminates.

Specimen ID	Flexural Strength (MPa)	Flexural Modulus (GPa)	Strain at Maximum Stress (%)
H_PP_F_L	85.9 ± 3.9	10.5 ± 0.4	2.6 ± 0.3
H_PP_B_L	101.0 ± 0.9	9.5 ± 0.1	3.2 ± 0.1
H_PPc_F_L	106.8 ± 2.3	12.6 ± 0.4	2.5 ± 0.3
H_PPc_B_L	129.3 ± 1.8	9.9 ± 0.4	3.2 ± 0.6
H_Ep_F_L	128.6 ± 4.8	14.4 ± 0.4	2.4 ± 0.2
H_Ep_B_L	165.2 ± 5.4	11.2 ± 0.2	1.8 ± 0.2

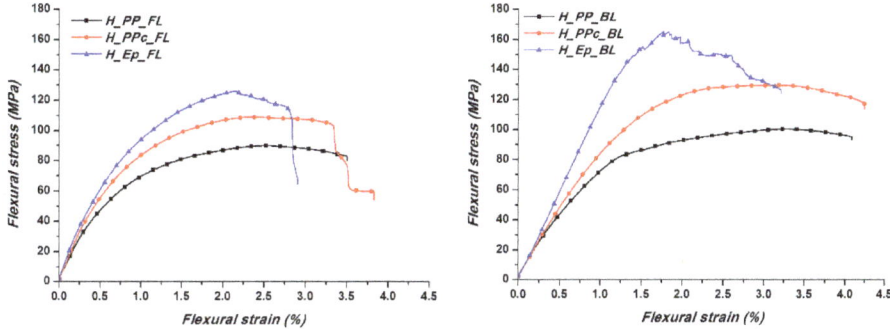

Figure 3. Typical stress vs. strain curves for flax/basalt composites along the warp (F_L) and weft (B_L) fiber directions.

For both matrices, a difference in the flexural modulus can be highlighted between warp and weft directions, which is likely to be ascribed to the higher fiber volume fraction of flax fibers in specimens when tested with flax fibers oriented in the longitudinal direction. As expected, composites with a thermoplastic matrix exhibited a much more ductile behavior compared to the epoxy resin. The ductility of PP was able to partially relieve the stress concentrations created in the matrix by defects (kink bands) always present in flax fibers [29]. This explanation supports the higher ductility observed in specimens tested along the basalt fibers, which are not characterized by such defects. Specimens tested with basalt fibers aligned along the longitudinal direction showed the highest flexural strength irrespective of the matrix type, which is due to the better absolute mechanical properties of basalt fibers compared to flax fibers. Usually flax fiber reinforced composites are characterized by a sudden failure due to the poor strain at failure of flax fibers (~1.2–1.8%) [30,31], while in the present case the hybridization with much more ductile basalt fibers [32] allowed the composites to fail in a much more gradual manner, especially when tested with basalt fibers in the longitudinal direction.

All the curves exhibited a significant non-linear behavior at low strains, which needs to be ascribed to the presence of flax fibers and seems only to be emphasized by the ductile PP. Many authors have pointed out this behavior [33,34] that seems to represent a peculiarity of natural fibers, as it has been reported to occur not only in flax [35,36] but also in wood [37] and hemp [38]. Recent studies tried to provide explanations for this behavior, and several mechanisms have been proposed, including cellulose microfibrils reorientation, shear strain-induced crystallization of the amorphous paracrystalline components and degree of ellipticity of the fiber's cross-section [38,39]. Indeed, this effect is significantly reduced in epoxy-based laminates when tested with basalt fibers in the longitudinal direction.

The presence of a coupling agent in the PP matrix caused an increase in both the flexural strength and modulus, without affecting in a significant way the ductility of the resulting composites. These results are due to the improvement of fiber/matrix interfacial adhesion for both fiber types. It is well known that the interactions between the anhydride groups of maleated coupling agents and the hydroxyl groups of flax and basalt fibers through esterification reactions can alleviate fiber/matrix incompatibility issues [40,41].

The fiber/matrix adhesion has been investigated by scanning electron microscopy. The level of interfacial adhesion of flax and basalt fibers with neat PP was found to be nonoptimal, as can be seen in Figure 4, where pull-out (Figure 4a,c,f) and debonding (Figure 4b,d,e) represent the dominant failure modes. In Figure 5 it is possible to note that both fibers, flax and basalt, benefited from the addition of the coupling agent. The pull-out was significantly reduced, and the fibers appear to be covered with a layer of polymer matrix, with ligaments connecting the fibers to the matrix (Figure 5a,b,d–f). For comparison purposes, the surface of pristine flax and basalt fibers is included in Figure 5g,h, respectively.

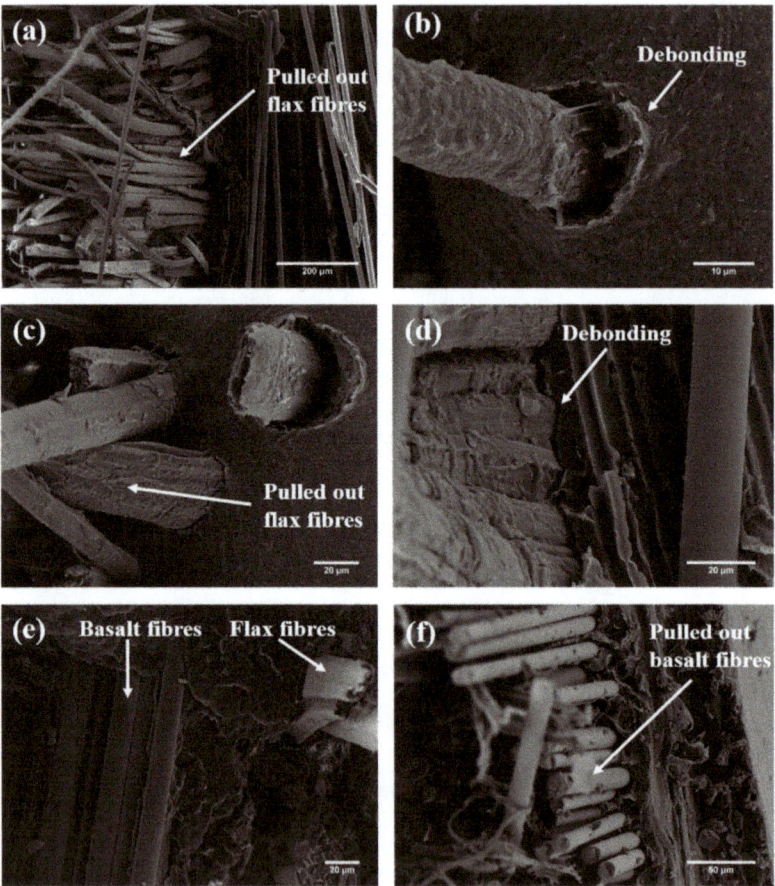

Figure 4. SEM micrographs of failed flexural H_PP specimens at different magnifications, highlighting the presence of extensive fiber pull-out (**a,c,f**) and fiber/matrix debonding (**b,d,e**).

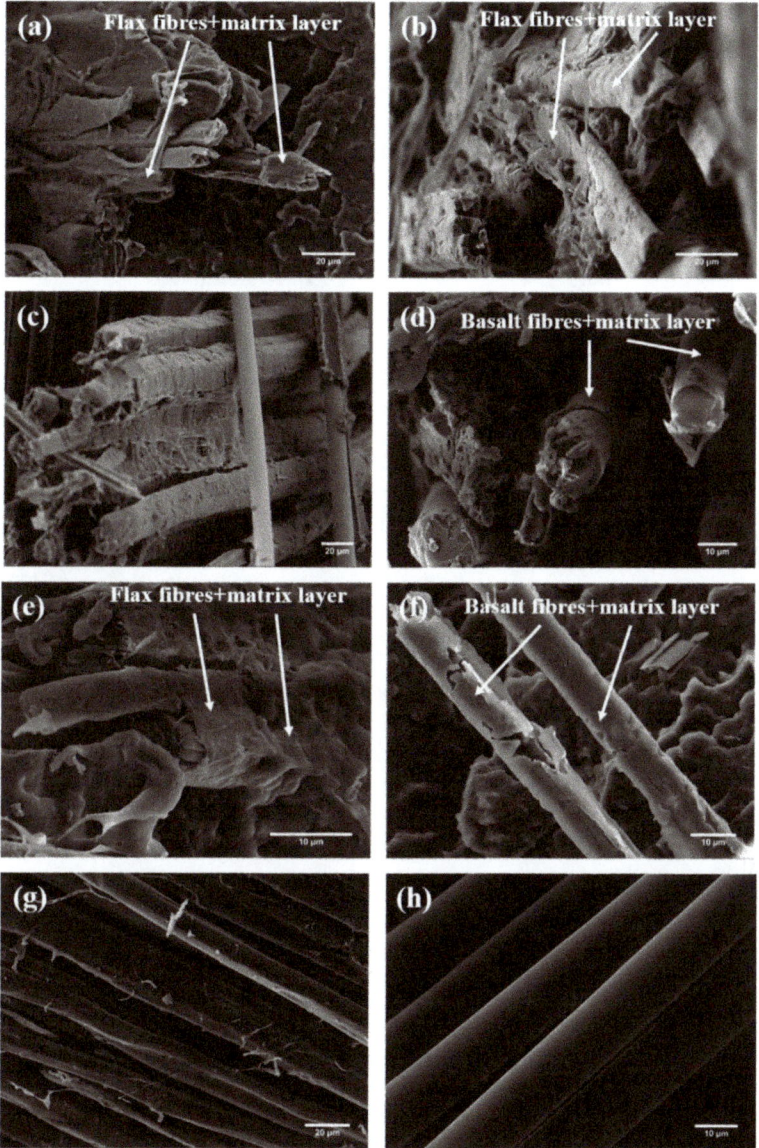

Figure 5. SEM micrographs of failed flexural H_PPc specimens at different magnifications, highlighting the presence of matrix layers on basalt (**c,d,f**) and flax fibers (**a,b,e**). Micrographs (**g**) and (**h**) show pristine flax and basalt fibers as extracted from the fabric, respectively.

For epoxy-based composites, the fracture surface exhibited a "blocky" appearance (Figure 6a,b,d) especially with basalt fibers fractured on the same plane, while also flax fibers were characterized by a sufficient level of adhesion as it can be inferred from the extensive fiber fibrillation (Figure 6c) coupled with the presence of epoxy residues on the fiber surface (Figure 6d,f). This supports the good mechanical behavior exhibited by the biocomposites, when compared with results available in literature.

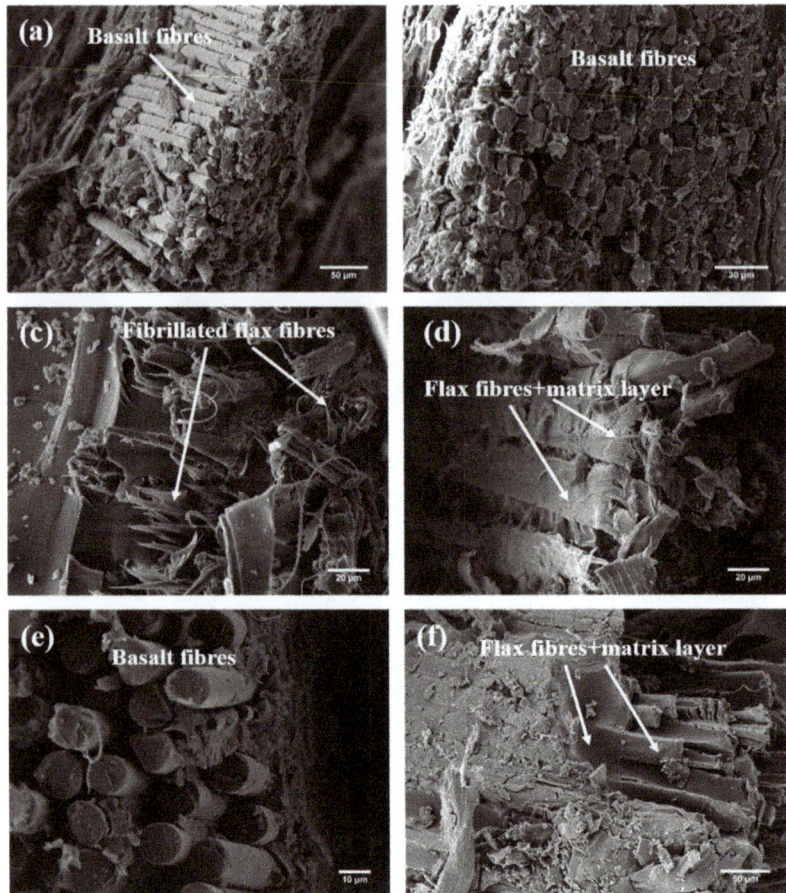

Figure 6. SEM micrographs of failed flexural H_Ep specimens at different magnifications, the presence of matrix layers on basalt (**a,b,e**) and flax fibers (**d,f**) and flax fiber fibrillation (**c**).

Despite the difficulties in comparing composites with different fiber volume fractions and fiber architectures, Meredith et al. [42] investigated several flax/epoxy composites based on the same weave style (Twill 2/2). The flexural strength was found to range from 57.0 to 195.2 MPa for fiber volume fractions from 37 to 54%, while flexural modulus was in the range 2.12–7.81 GPa. Goutianos et al. [43] compared different flax fiber architectures in a vinylester matrix for fiber volume fractions in the range of 29 to 35%. Composites showed flexural strengths ranging from 80 to 140 MPa and flexural moduli from 5 to 10 GPa. Cihan et al. [44] investigated the mechanical performance of woven flax/E-glass hybrid composites. For flax fiber reinforced composites, with a fiber volume fraction around 0.36, the tensile strength and stiffness were reported to be 86.43 MPa and 8.89 GPa, respectively, while Blanchard et al. [45] for a volume fraction of flax fibers equal to 0.37 in an epoxy matrix, found a flexural strength and stiffness of 114.91 MPa and 6.13 GPa, respectively. In the present case, by adding around 13% by volume of basalt fibers, it is possible to increase the flexural strength and modulus of 43% and 83%, respectively. In particular, it is worth noting the increase in flexural stiffness that can be obtained with the present hybrid, which could represent an opportunity to introduce flax fibers in semi-structural applications.

3.2. Low-Velocity Impact Behaviour

The representative force vs displacement curves of the hybrid composites as a function of matrix type and impact energy are reported in Figure 7. Matrix effect was not significant at 5 J (Figure 7a) as all the samples exhibited similar hysteresis loops, but for a 10 J-impact event some differences can be highlighted. In particular, epoxy-based laminates showed a more drastic load drop (Figure 7b) followed by the unloading curve. This substantial load drop points toward a loss of elastic energy and the vulnerability to fiber breakage and subsequent perforation, which was indeed reached at 15 J for thermoset-based composites (Figure 7c).

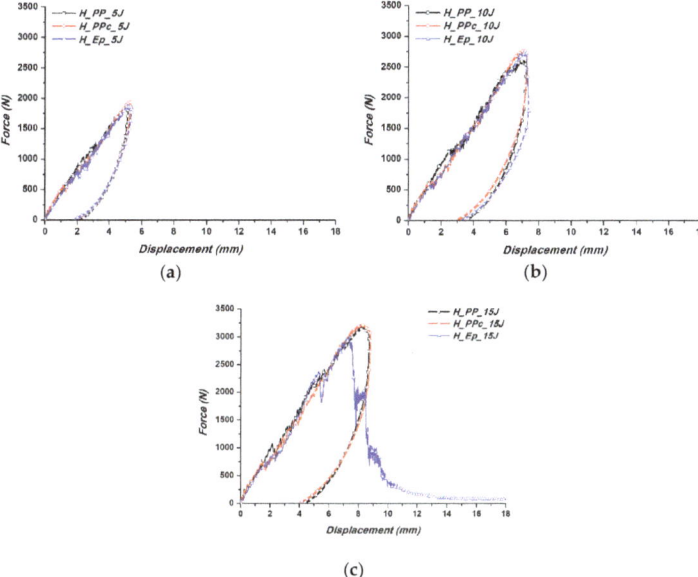

Figure 7. Typical force–displacement curves of flax/basalt hybrid composites at various impact energy levels: (**a**) 5 J; (**b**) 10 J, and (**c**) 15 J.

Epoxy laminates presented the lowest impact resistance at all impact energies, whereas thermoplastic-based laminates exhibited a very similar impact damage behavior in terms of force vs displacement response. The results compare quite favorably with those available in the literature for flax fiber reinforced epoxy composites. Bensadoun et al. [46] investigated the impact behavior of laminates with different flax fiber architectures and matrix types. Despite the unavoidable differences in impact parameters, similar conditions were used for laminates with a thickness of 2 mm and a total fiber volume fraction of 0.40. The authors found perforation energies ranging from 5.7 J to 7 J, which are lower than the value found in the present case (15 J).

It is suggested that basalt fiber hybridization in intraply configuration can improve the common poor impact response reported in the literature of plant fiber composites. The shift from a thermoset to a thermoplastic matrix markedly influenced the absorbed energy, as it can be clearly observed in Figure 8, where 30 J and 20 J were absorbed at perforation by PP- and PPc-based laminates, respectively.

In perforation impact tests, the fibers need to be broken to achieve perforation. Therefore, the stronger and tougher the fibers, the more pronounced is the role played by the fiber fracture energy in the total perforation energy. In the present case, as the fibers are the same for all the laminates, the relative contribution of the matrix ductility to the composite impact behavior is more important. Hybrid epoxy composites, at each impact energy, showed the highest damage degree, i.e., the ratio of absorbed energy to the kinetic impact energy, while the energy dissipated during impact is almost the same in thermoplastic-based laminates at least up to 15 J. This ratio increases up to a maximum value of unity when perforation occurs.

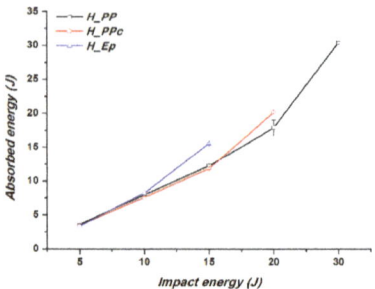

Figure 8. Impact energy vs. absorbed energy curves at various impact energy levels and matrix type.

The higher ductility of the PP matrix allows more energy absorption through plastic deformation compared to the brittle epoxy matrix, as observed by other authors [25,46]. This is supported by the assessment of the residual indentation depth, also known as dent depth, which has been performed by laser profilometry after impact (Figure 9).

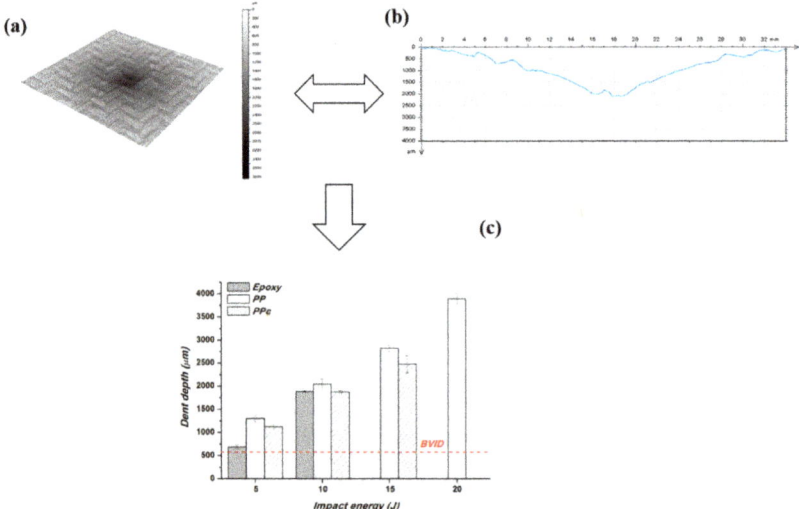

Figure 9. Dent depth: (**a**) 3-D image of damaged specimen; (**b**) extracted profile and (**c**) changes in dent depth on the impacted face of specimens as a function of impact energy and matrix type.

This type of damage can involve local matrix failure, local matrix plastic deformation, and local fiber breakage. At all impact energies, composites based on neat polypropylene exhibited a higher residual plastic deformation. It is also worth noting that, in accordance with the commonly accepted definition of BVID (Barely Visible Impact Damage) [25], the threshold of 0.6 mm (highlighted in Figure 9) is reached for all composites already at 5 J.

To better appreciate differences between compatibilized and neat polypropylene matrices, the representative force vs displacement curves for the whole range of impact energies are given in Figure 10. The effect of matrix type becomes significant with increasing impact energy (>15 J) (Figure 10). At 20 J, a more drastic load drop was exhibited by the compatibilized PP, while an energy of 30 J was needed to perforate the specimens based on neat PP, likely due to different damage absorption mechanisms associated with the different extent of fiber/matrix adhesion.

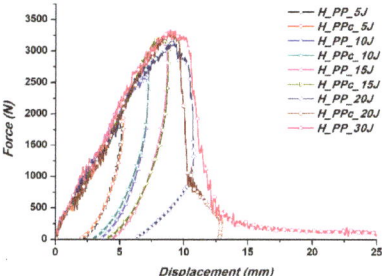

Figure 10. Typical force–displacement curves of flax/basalt hybrid thermoplastic composites in the whole range of impact energies.

Macroscopic observation of the impacted specimens corroborates these results (Figures 11–13). The scale in each photograph has been adapted to highlight the different damage mechanisms. Thermoset composites exhibited a sharp failure pattern, typical of their brittle character (Figure 11), with the presence of cracks along warp/weft directions in the impacted side already for an impact energy level as low as 5 J. The intraply hybridization prevented the samples from exhibiting the characteristic diamond-shaped fracture surface on the rear side [15,25,46], which was characterized by fiber splitting and cracks mainly along the weft direction. It is interesting to note that these cracks run parallel to the basalt yarns and caused the failure only of the weaker flax yarns. After perforation at 15 J, a limited fiber pull-out was detected, thus, confirming the sufficient level of fiber/matrix adhesion.

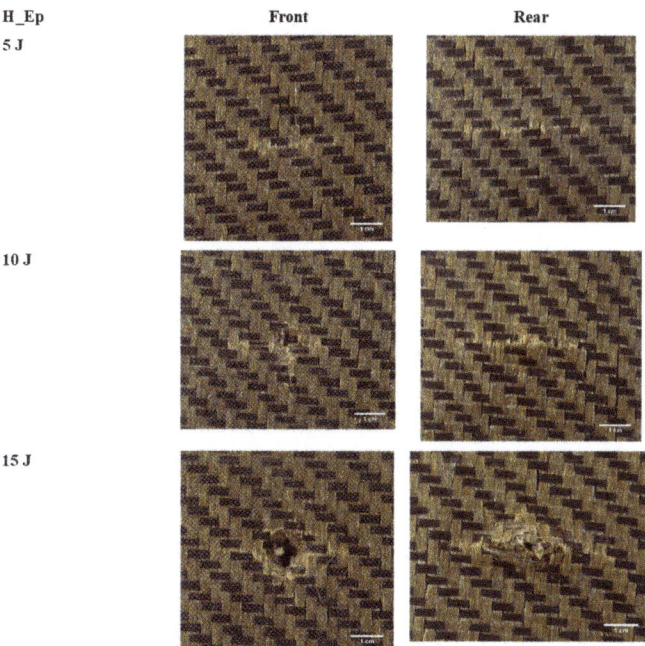

Figure 11. Close-up view of impact damage progression on front and rear surfaces of epoxy-based impacted specimens.

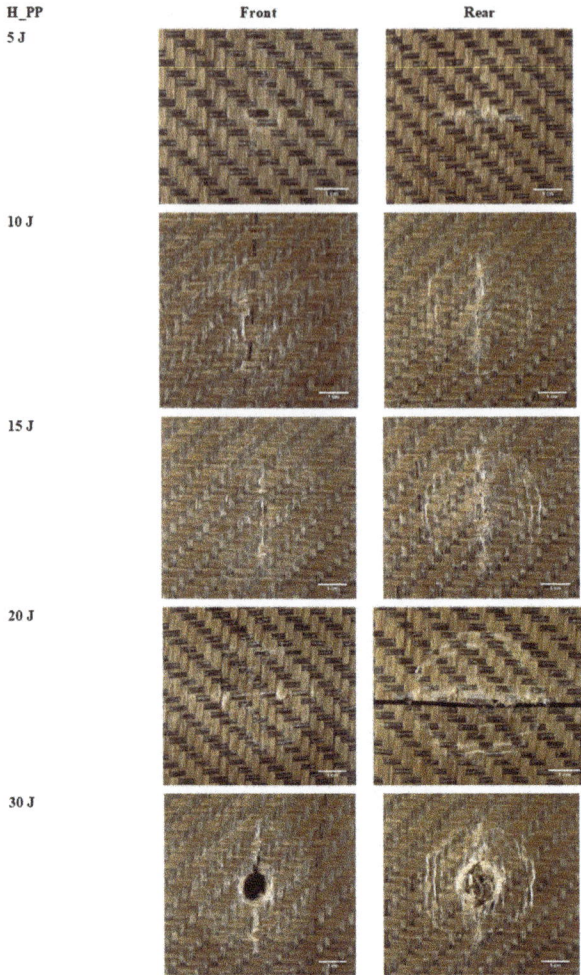

Figure 12. Close-up view of impact damage progression on front and rear surfaces of PP-based impacted specimens.

Thermoplastic-based laminates showed a different behavior, with a more ductile response to impact loading. In both cases (Figures 12 and 13) an extensive plastic deformation and a wider damaged area can be easily observed with the presence of matrix cracks and flax fiber failures on the impacted side.

A significant difference between compatibilized and neat PP is represented by the extensive stress whitening that occurred in all compatibilized samples (Figure 13). This phenomenon increased with increasing impact energy and extended up to the edges of the impacted plates. This can be related to the better fiber/matrix interfacial adhesion induced by the coupling agent that hindered the plastic deformation of the matrix, thus, forcing a larger amount of the sample to respond to the impact loading. This behavior was not observed in neat PP-based composites where the matrix was free to plastically deform under the contact point with the impactor, thus, causing a deeper indentation (Figure 9) and a higher energy absorption (Figure 8).

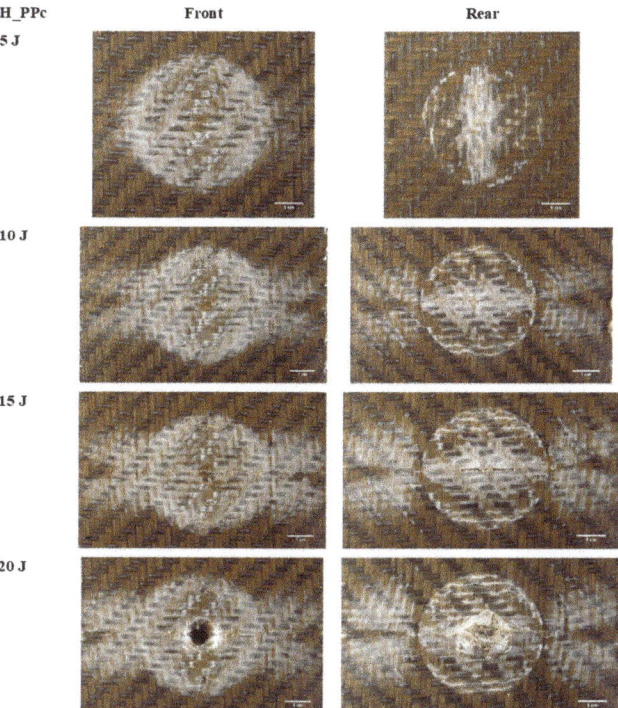

Figure 13. Close-up view of impact damage progression on front and rear surfaces of PPc-based impacted specimens.

It is worth mentioning that no necking and/or wrinkling were detected as damage mechanisms in polypropylene-based composites, which need to be avoided as they absorb additional energy compared to properly tested samples, as usually found in self-reinforced polypropylene composites if tested with small geometry ratios, i.e., the ratio of the sample size divided by the clamp size [47].

The higher bending strength of the PPc-based laminates combined with the higher strain at break of the thermoplastic compared to the thermoset matrix delayed the onset of first damage because plastic deformation occurred before the matrix failure, thus, ensuring a higher peak force under impact, which represents the maximum load a laminate can tolerate before major damage (Figure 14).

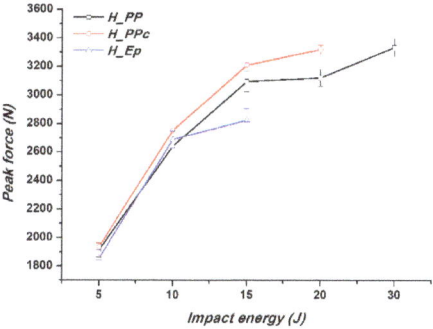

Figure 14. Peak force vs kinetic impact energy for flax/basalt hybrid composites as a function of matrix type.

4. Conclusions

A new intraply hybrid composite using basalt and flax fibers has been manufactured and tested in both quasi-static (three-point bending) and dynamic (low-velocity impact) conditions with the main aim of assessing the effect of matrix behavior. In this regard, three different matrices have been investigated, namely epoxy, polypropylene, and a polypropylene compatibilized with maleic anhydride. The matrix ductility was found to have a significant influence on the impact response only at energies higher than 5 J, while composites based on compatibilized PP showed the best combination of properties in terms of quasi-static strength, energy absorption, peak force, and perforation energy.

Basalt fiber hybridization in intraply configuration significantly improved the common poor impact response of plant fiber composites usually reported in the literature, preventing the growth of cracks in a diamond-shaped pattern and balancing the poor transverse strength of flax fibers. Thermoplastic-based laminates exhibited a concentrated (circular) damaged zone at perforation due to the extensive plastic deformation that hindered further extension of the cross-shaped cracks, thus, enhancing the energy absorption of the resulting hybrid laminates. For 2-mm thick hybrid laminates, perforation threshold was found to lie in the range of 20 to 30 J, depending on the presence or not of the coupling agent, respectively. The results confirm the suitability of polypropylene-based intraply hybrid composites to impact-related applications.

Author Contributions: Conceptualization, F.S., J.T., C.S. and P.R.; methodology, F.S. and L.F.; formal analysis, F.S., J.T. and C.S.; investigation, L.F., G.S., F.C., M.R.R. and V.A.; writing—original draft preparation, F.S.; writing—review and editing, F.S., P.R., M.R.R., V.A. and J.T.; supervision, F.S. and J.T.

Funding: This research received no external funding.

Conflicts of Interest: The authors declare no conflict of interest.

References

1. Swolfs, Y.; Verpoest, I.; Gorbatikh, L. Recent advances in fibre-hybrid composites: Materials selection, opportunities and applications. *Int. Mater. Rev.* **2018**, 1–35. [CrossRef]
2. Swolfs, Y.; Gorbatikh, L.; Verpoest, I. Fibre hybridisation in polymer composites: A review. *Compos. Part A Appl. Sci. Manuf.* **2014**, *67*, 181–200. [CrossRef]
3. Manders, P.W.; Bader, M.G. The strength of hybrid glass/carbon fibre composites. *J. Mater. Sci.* **1981**, *16*, 2233–2245. [CrossRef]
4. Safri, S.N.A.; Sultan, M.T.H.; Jawaid, M.; Jayakrishna, K. Impact behaviour of hybrid composites for structural applications: A review. *Compos. Part B Eng.* **2018**, *133*, 112–121. [CrossRef]
5. Naik, N.; Ramasimha, R.; Arya, H.; Prabhu, S.; ShamaRao, N. Impact response and damage tolerance characteristics of glass–carbon/epoxy hybrid composite plates. *Compos. Part B Eng.* **2001**, *32*, 565–574. [CrossRef]
6. Hosur, M.V.; Adbullah, M.; Jeelani, S. Studies on the low-velocity impact response of woven hybrid composites. *Compos. Struct.* **2005**, *67*, 253–262. [CrossRef]
7. Sayer, M.; Bektaş, N.B.; Sayman, O. An experimental investigation on the impact behavior of hybrid composite plates. *Compos. Struct.* **2010**, *92*, 1256–1262. [CrossRef]
8. Gustin, J.; Joneson, A.; Mahinfalah, M.; Stone, J. Low velocity impact of combination Kevlar/carbon fiber sandwich composites. *Compos. Struct.* **2005**, *69*, 396–406. [CrossRef]
9. Salehi-Khojin, A.; Mahinfalah, M.; Bashirzadeh, R.; Freeman, B. Temperature effects on Kevlar/hybrid and carbon fiber composite sandwiches under impact loading. *Compos. Struct.* **2007**, *78*, 197–206. [CrossRef]
10. Wan, Y.Z.; Wang, Y.L.; He, F.; Huang, Y.; Jiang, H.J. Mechanical performance of hybrid bismaleimide composites reinforced with three-dimensional braided carbon and Kevlar fabrics. *Compos. Part A Appl. Sci. Manuf.* **2007**, *38*, 495–504. [CrossRef]
11. Imielińska, K.; Castaings, M.; Wojtyra, R.; Haras, J.; Le Clezio, E.; Hosten, B. Air-coupled ultrasonic C-scan technique in impact response testing of carbon fibre and hybrid: Glass, carbon and Kevlar/epoxy composites. *J. Mater. Process. Technol.* **2004**, *157–158*, 513–522. [CrossRef]

12. Ying, S.; Mengyun, T.; Zhijun, R.; Baohui, S.; Li, C. An experimental investigation on the low-velocity impact response of carbon–aramid/epoxy hybrid composite laminates. *J. Reinf. Plast. Compos.* **2017**, *36*, 422–434. [CrossRef]
13. Petrucci, R.; Santulli, C.; Puglia, D.; Nisini, E.; Sarasini, F.; Tirillò, J.; Torre, L.; Minak, G.; Kenny, J.M. Impact and post-impact damage characterisation of hybrid composite laminates based on basalt fibres in combination with flax, hemp and glass fibres manufactured by vacuum infusion. *Compos. Part B Eng.* **2015**, *69*, 507–515. [CrossRef]
14. Ahmed, K.S.; Vijayarangan, S.; Kumar, A. Low Velocity Impact Damage Characterization of Woven Jute Glass Fabric Reinforced Isothalic Polyester Hybrid Composites. *J. Reinf. Plast. Compos.* **2007**, *26*, 959–976. [CrossRef]
15. Sarasini, F.; Tirillò, J.; D'Altilia, S.; Valente, T.; Santulli, C.; Touchard, F.; Chocinski-Arnault, L.; Mellier, D.; Lampani, L.; Gaudenzi, P. Damage tolerance of carbon/flax hybrid composites subjected to low velocity impact. *Compos. Part B Eng.* **2016**, *91*, 144–153. [CrossRef]
16. Al-Hajaj, Z.; Sy, B.L.; Bougherara, H.; Zdero, R. Impact properties of a new hybrid composite material made from woven carbon fibres plus flax fibres in an epoxy matrix. *Compos. Struct.* **2019**, *208*, 346–356. [CrossRef]
17. Förster, T.; Hao, B.; Mäder, E.; Simon, F.; Wölfel, E.; Ma, P.-C.; Förster, T.; Hao, B.; Mäder, E.; Simon, F.; et al. CVD-Grown CNTs on Basalt Fiber Surfaces for Multifunctional Composite Interphases. *Fibers* **2016**, *4*, 28. [CrossRef]
18. Fragassa, C.; Pavlovic, A.; Santulli, C. Mechanical and impact characterisation of flax and basalt fibre vinylester composites and their hybrids. *Compos. Part B Eng.* **2018**, *137*, 247–259. [CrossRef]
19. Dhakal, H.N.; Sarasini, F.; Santulli, C.; Tirillò, J.; Zhang, Z.; Arumugam, V. Effect of basalt fibre hybridisation on post-impact mechanical behaviour of hemp fibre reinforced composites. *Compos. Part A Appl. Sci. Manuf.* **2015**, *75*, 54–67. [CrossRef]
20. Papa, I.; Ricciardi, M.R.; Antonucci, V.; Pagliarulo, V.; Lopresto, V. Impact behaviour of hybrid basalt/flax twill laminates. *Compos. Part B Eng.* **2018**, *153*, 17–25. [CrossRef]
21. Pegoretti, A.; Fabbri, E.; Migliaresi, C.; Pilati, F. Intraply and interply hybrid composites based on E-glass and poly(vinyl alcohol) woven fabrics: Tensile and impact properties. *Polym. Int.* **2004**, *53*, 1290–1297. [CrossRef]
22. Wang, X.; Hu, B.; Feng, Y.; Liang, F.; Mo, J.; Xiong, J.; Qiu, Y. Low velocity impact properties of 3D woven basalt/aramid hybrid composites. *Compos. Sci. Technol.* **2008**, *68*, 444–450. [CrossRef]
23. Zhang, C.; Rao, Y.; Li, Z.; Li, W. Low-Velocity Impact Behavior of Interlayer/Intralayer Hybrid Composites Based on Carbon and Glass Non-Crimp Fabric. *Materials* **2018**, *11*, 2472. [CrossRef] [PubMed]
24. Bandaru, A.K.; Patel, S.; Sachan, Y.; Alagirusamy, R.; Bhatnagar, N.; Ahmad, S. Low velocity impact response of 3D angle-interlock Kevlar/basalt reinforced polypropylene composites. *Mater. Des.* **2016**, *105*, 323–332. [CrossRef]
25. Vieille, B.; Casado, V.M.; Bouvet, C. About the impact behavior of woven-ply carbon fiber-reinforced thermoplastic- and thermosetting-composites: A comparative study. *Compos. Struct.* **2013**, *101*, 9–21. [CrossRef]
26. Arikan, V.; Sayman, O. Comparative study on repeated impact response of E-glass fiber reinforced polypropylene & epoxy matrix composites. *Compos. Part B Eng.* **2015**, *83*, 1–6. [CrossRef]
27. Mutjé, P.; Vallejos, M.E.; Gironès, J.; Vilaseca, F.; López, A.; López, J.P.; Méndez, J.A. Effect of maleated polypropylene as coupling agent for polypropylene composites reinforced with hemp strands. *J. Appl. Polym. Sci.* **2006**, *102*, 833–840. [CrossRef]
28. Mohanty, A.; Misra, M.; Hinrichsen, G. Biofibres, biodegradable polymers and biocomposites: An overview. *Macromol. Mater. Eng.* **2000**, *276–277*, 1–24. [CrossRef]
29. Hughes, M.; Carpenter, J.; Hill, C. Deformation and fracture behaviour of flax fibre reinforced thermosetting polymer matrix composites. *J. Mater. Sci.* **2007**, *42*, 2499–2511. [CrossRef]
30. Audibert, C.; Andreani, A.-S.; Lainé, É.; Grandidier, J.-C. Mechanical characterization and damage mechanism of a new flax-Kevlar hybrid/epoxy composite. *Compos. Struct.* **2018**, *195*, 126–135. [CrossRef]
31. Wambua, P.; Ivens, J.; Verpoest, I. Natural fibres: Can they replace glass in fibre reinforced plastics? *Compos. Sci. Technol.* **2003**, *63*, 1259–1264. [CrossRef]
32. Fiore, V.; Scalici, T.; Di Bella, G.; Valenza, A. A review on basalt fibre and its composites. *Compos. Part B Eng.* **2015**, *74*, 74–94. [CrossRef]

33. Poilane, C.; Cherif, Z.E.; Richard, F.; Vivet, A.; Ben Doudou, B.; Chen, J. Polymer reinforced by flax fibres as a viscoelastoplastic material. *Compos. Struct.* **2014**, *112*, 100–112. [CrossRef]
34. Hughes, M.; Hill, C.A.S.; Hague, J.R.B. The fracture toughness of bast fibre reinforced polyester composites Part 1 Evaluation and analysis. *J. Mater. Sci.* **2002**, *37*, 4669–4676. [CrossRef]
35. Baley, C. Analysis of the flax fibres tensile behaviour and analysis of the tensile stiffness increase. *Compos. Part A Appl. Sci. Manuf.* **2002**, *33*, 939–948. [CrossRef]
36. Mahboob, Z.; El Sawi, I.; Zdero, R.; Fawaz, Z.; Bougherara, H. Tensile and compressive damaged response in Flax fibre reinforced epoxy composites. *Compos. Part A Appl. Sci. Manuf.* **2017**, *92*, 118–133. [CrossRef]
37. Navi, P.; Rastogi, P.; Gresse, V.; Tolou, A. Micromechanics of wood subjected to axial tension. *Wood Sci. Technol.* **1995**, *29*, 411–429. [CrossRef]
38. Placet, V.; Cissé, O.; Lamine Boubakar, M. Nonlinear tensile behaviour of elementary hemp fibres. Part I: Investigation of the possible origins using repeated progressive loading with in situ microscopic observations. *Compos. Part A Appl. Sci. Manuf.* **2014**, *56*, 319–327. [CrossRef]
39. Del Masto, A.; Trivaudey, F.; Guicheret-Retel, V.; Placet, V.; Boubakar, L. Nonlinear tensile behaviour of elementary hemp fibres: A numerical investigation of the relationships between 3D geometry and tensile behaviour. *J. Mater. Sci.* **2017**, *52*, 6591–6610. [CrossRef]
40. Keener, T.; Stuart, R.; Brown, T. Maleated coupling agents for natural fibre composites. *Compos. Part A Appl. Sci. Manuf.* **2004**, *35*, 357–362. [CrossRef]
41. Matuana, L.M.; Balatinecz, J.J.; Sodhi, R.N.S.; Park, C.B. Surface characterization of esterified cellulosic fibers by XPS and FTIR Spectroscopy. *Wood Sci. Technol.* **2001**, *35*, 191–201. [CrossRef]
42. Meredith, J.; Coles, S.R.; Powe, R.; Collings, E.; Cozien-Cazuc, S.; Weager, B.; Müssig, J.; Kirwan, K. On the static and dynamic properties of flax and Cordenka epoxy composites. *Compos. Sci. Technol.* **2013**, *80*, 31–38. [CrossRef]
43. Goutianos, S.; Peijs, T.; Nystrom, B.; Skrifvars, M. Development of Flax Fibre based Textile Reinforcements for Composite Applications. *Appl. Compos. Mater.* **2006**, *13*, 199–215. [CrossRef]
44. Cihan, M.; Sobey, A.J.; Blake, J.I.R. Mechanical and dynamic performance of woven flax/E-glass hybrid composites. *Compos. Sci. Technol.* **2019**, *172*, 36–42. [CrossRef]
45. Blanchard, J.M.F.A.; Sobey, A.J.; Blake, J.I.R. Multi-scale investigation into the mechanical behaviour of flax in yarn, cloth and laminate form. *Compos. Part B Eng.* **2016**, *84*, 228–235. [CrossRef]
46. Bensadoun, F.; Depuydt, D.; Baets, J.; Verpoest, I.; van Vuure, A.W. Low velocity impact properties of flax composites. *Compos. Struct.* **2017**, *176*, 933–944. [CrossRef]
47. Meerten, Y.; Swolfs, Y.; Baets, J.; Gorbatikh, L.; Verpoest, I. Penetration impact testing of self-reinforced composites. *Compos. Part A Appl. Sci. Manuf.* **2015**, *68*, 289–295. [CrossRef]

© 2019 by the authors. Licensee MDPI, Basel, Switzerland. This article is an open access article distributed under the terms and conditions of the Creative Commons Attribution (CC BY) license (http://creativecommons.org/licenses/by/4.0/).

Article

Mechanical, Degradation and Water Uptake Properties of Fabric Reinforced Polypropylene Based Composites: Effect of Alkali on Composites

Mohammad Bellal Hoque [1,*], Solaiman [2], A.B.M. Hafizul Alam [3], Hasan Mahmud [2] and Asiqun Nobi [1]

1. College of Fashion Technology & Management, Uttara, Dhaka 1230, Bangladesh; rimonashiq@gmail.com
2. Textile Engineering College, Begumganj, Noakhali 3820, Bangladesh; solaiman1@btec.com (S.); hasanmahmud@btec.com (H.M.)
3. Department of Textile Engineering, World University of Bangladesh, Green Road, Dhaka 1209, Bangladesh; hafizul1@textiles.wub.edu.bd
* Correspondence: bellal.te@fiu.edu.bd; Tel.: +880-1911-427230

Received: 16 August 2018; Accepted: 12 November 2018; Published: 6 December 2018

Abstract: In this study, a fabric was manufactured consisting of 50% pineapple, 25% jute and 25% cotton fibers by weight, to make composites using polypropylene (PP) as a matrix material. We used compression molding technique, which kept 30% of the fabric content by total weight as the composite. The tensile strength (TS), tensile modulus (TM), elongation break (Eb%), bending strength (BS) and bending modulus (BM) were investigated. From analyzed data, it was found that the composite values of TS, TM, Eb%, BS and BM were 58 MPa, 867 MPa, 22.38%, 42 MPa and 495 MPa, respectively. The TS, TM, Eb%, BS and BM of the neat polypropylene sheet were 28 MPa, 338 MPa, 75%, 20 MPa and 230 MPa, respectively. Due to fabric reinforcement, composite values for TS, TM, BS and BM increased 107%, 156%, 110% and 115%, respectively in comparison with a polypropylene sheet. A water absorption test was performed by dipping the composite samples in deionized water and it was noticed that water absorption was lower for PP-based composites. For investigating the effect of alkali, we sunk the composites in a solution containing 3%, 5% and 7% sodium hydroxide alkali solutions by weight, for 60 min after which their mechanical properties were investigated. A degradation test was carried out by putting the samples in soil for six months and it was noticed that the mechanical properties of fabric/PP composites degraded slowly.

Keywords: polypropylene (PP); composites; natural fiber; fabric; compression molding

1. Introduction

Fabricating composites by using natural fibers has been of great interest because natural fiber is biodegradable and environment friendly [1–14]. Composites are inexpensive and also not harmful to the environment. The advantages of fiber reinforced materials (FRM) over unreinforced ones are well known, and their characteristics are useful in many fields, and as such, they are used for many applications. FRM are widely used in aerospace [15] and construction [16] applications. Moreover net [17] or long [18] continuous fiber reinforced composites are used in medical science. Nowadays, the most used fibers to reinforce composites are synthetic, but the results obtained with natural fibers are promising [19]. Synthetic fiber reinforced polymers are expensive and have an environmental impact. Natural fibers which are cellulose-based are replacing synthetic fibers as they resolve this issue. This is the reason that the interest of using natural fibers in combination with thermoplastic material is increasing day by day, especially for high volume and low cost applications. Natural fibers have inherent properties of polarity and hydrophilicity, which can be removed by using non-polar thermoplastic material. These properties are not considered when they are not harmful to the environment [20–23].

Scientists and engineers are paying more attention to replacing synthetic fibers with natural fibers that have analogous physical and mechanical properties. Various other matters should be kept in consideration while choosing raw materials such as cost, environmental impact, hygiene, flexibility, ease of collection and be availability, which are directly related to the suitability of natural fibers [24–27]. Being a renewable resource, natural fibers provide a long lasting solution, beside the economical and hygienic benefits. Moreover, the processing of natural fibers is cost effective and provides good mechanical and physical properties [28–33].

Pineapple leaf fiber (PALF) is a waste material in the agriculture sector which is cultivated widely in Asia across the world. Very few tropical fruits are as essential as pineapple (*Ananas comosus*). Pineapple fruit has an important commercial value, and the leaves of pineapple are waste material which yields natural fibers [34]. It is chemically composed of 70–82% holocellulose, 5–12% lignin and 1.1% ash. It has excellent mechanical properties due to the higher percentage of hollocellulose. This is why it can be implemented in fabricating of reinforced polymer composites [35,36]. Cotton is an important natural fiber and produced in many parts of the Asian continent such as China, India, Pakistan, Bangladesh etc. Among these countries, China is the biggest cotton producer. Cotton fiber is comprised of 94% cellulose. There are various factors on which the strength of cotton fiber is dependent like fiber structure, microfibril orientation, molecular weight of cellulose chains, the crystalline structure perfection and the convolution angle of microfibrils [37]. Another important natural fiber is jute fiber which is mostly produced in Bangladesh, India, China, Uzbekistan, Bhutan, Vietnam and Thailand. About 93% of the world's jute fiber is produced in Bangladesh and India. Clothes, ropes, bags, floor mats etc., are made from jute fiber. Moreover, it can be used as a good reinforcing agent with hydrophobic matrix material like polypropylene, polyethylene, low density polyethylene, and unsaturated polyester resin, etc. It provides advantages such as being light weight, cost effective, low density. Also it has having high availability, a high tensile modulus and low elongation at break [3,38]. Jute fibers are comprised of 82–85% hollocellulose, of which 58–63% is alpha-cellulose, which is responsible for the excellent mechanical properties of jute fiber [8]. For these reasons, to fabricate a lightweight and inexpensive composite, pineapple leaf, jute and cotton fibers were selected.

The most vital part of a composite are the matrix materials. As a matrix material, polypropylene has been extensively used with natural fibers in composite preparation [2–8]. Polypropylene (PP) is an amorphous thermoplastic polymer and is extensively used as an engineering thermoplastic material for its various important characteristics such as its dimensional stability, transparency, high heat distortion temperature, flame resistance and high impact strength. PP can also be used for filling, reinforcing and blending. PP composites with natural fiber are becoming more promising each day [9,39–44].

In this study, the mechanical and degradation properties of a fabric reinforced PP-based partially biodegradable composite were evaluated. Water uptake profiles and the effect of alkali on composite were also investigated.

2. Materials and Methods

2.1. Materials

A plain structure bleached fabric (consisting of 50% pineapple, 25% jute and 25% cotton) was made at the Bangladesh Jute Research Institute (BJRI), Dhaka, Bangladesh. The thread count in the vertical (warp) and horizontal (weft) directions of the fabric were 52 and 40 in one square inch. Figure 1 shows the weaved fabric. The granules of PP were bought from Polyolefin Company Limited, Singapore. The symbolic expression of PP is $(C_3H_6)_n$. Alkali (NaOH) was purchased from the local market of Dhaka, Bangladesh.

2.2. Fabrication of Fabric-Reinforced PP-Based Composites

The granules of polypropylene were placed inside of two plates of a heat press machine (Carver, INC, USA Model 3856) for making PP sheets. The heat press machine was used at a temperature of 180 °C for 5 min and a load of 2000 kg. The sheets were cooled inside the heat press machine. For composite fabrication, the prepared PP sheet and fabric were cut into the desired size. Composites were fabricated by sandwiching fabric between two sheets of PP and using the method as discussed for making PP sheets [2]. The fabric percentage of the composites was 30% by weight. The prepared composites were then packed in polyethylene bags.

Figure 1. Fabric.

2.3. Mechanical Properties of Composites

The tensile and bending properties of the composites were measured according to the European standard (ISO/DIS 527-1:2010) by using the Hounsfield series S testing machine (UK) with an initial separation of clamp of 20 mm and a loading force of 10 mm/min. The dimensions of fabric/PP composites were (60 × 10 × 1.60) mm^3. The test samples were conditioned at 25 °C and 50% relative humidity for three days prior to testing. The mechanical tests were carried out according to the vertical (warp) direction. The average of the results for at least five samples was taken for all the test values.

2.4. Water Absorption Profile of the Composites

Water absorption test of the fabric/PP composite was performed according to ASTM D-570. Water absorption tests were carried out on three samples (sample-1, sample-2 and sample-3) of the fabric/PP composite (Figure 2). Samples of the composite were weighed and dipped in beakers containing 500 mL of deionized water at room temperature (25 °C) for 1 h. Then after the time interval the samples were brought out of the beaker, wiped by using tissue paper and re-weighed. In this case, it showed no uptake after 40 min, so we carried out the test up to 1 h [14]. The water absorption percentage was determined by:

$$\text{Water absorption (\%)} = [(\text{Wet weight} - \text{Dry weight})/\text{Dry weight}] \times 100 \quad (1)$$

Figure 2. (a) Universal testing machine (UTM) and (b) water uptake (%) test.

2.5. Effect of Alkali

The alkali treatment was performed to investigate the effect of alkali on the composites. The composites were treated with aqueous solution of 3%, 5% and 7% sodium hydroxide (NaOH) for 24 h [2]. The samples were taken out of the solution after 24 h and washed in water to remove the remaining sodium hydroxide. Then the samples were dried at 70 °C for 1 h and their mechanical properties were evaluated.

2.6. Soil Degradation Test of the Composites

Soil degradation tests of the composite provided information on the mechanical properties of the composites retained after a certain time period of contact with. The degradation test of the fabric/PP composites was conducted for up to 24 weeks in soil. Composite samples were placed in soil for different periods of time. After a three week interval, samples were brought out carefully. Then cleaned with distilled water and dehydrated at 105 °C for 6 h, reposed at room temperature (25 °C) for 24 h and then the tensile and bending properties were measured [8].

3. Results and Discussion

3.1. Mechanical Properties of Composites

The tensile and bending property values of the PP matrix and fabric/PP composites are represented in Tables 1 and 2. From the tables, the tensile strength (TS), tensile modulus (TM), elongation at break (Eb%), bending strength (BS) and bending modulus (BM) of the PP sheet were found to be 28 MPa, 338 MPa, 75%, 20 MPa and 230 MPa respectively. TS, TM, BS and BM for the fabric/PP composites were found to be 58, 867, 42 and 483 MPa respectively. Pineapple leaf, cotton & jute fiber based PP composites gained a 107% increase in TS and a 110% increase in BS over that of the matrix PP. It was also observed that TM and BM improved by 156% and 115% respectively over that of the matrix material PP. By analyzing the values, it was found that 30 wt% fabric/PP composite exhibited improved mechanical properties (TS, TM, BS and BM). But Eb% was reduced drastically compared to PP. Motaleb et al. stated that the TS, TM, Eb%, BS and BM of PALF/PP composite (30% fiber content) were 61 MPa, 1096 MPa, 14.05%, 31 MPa and 420 MPa, respectively [14]. Compared with the present study, reduced mechanical properties were observed due to the addition of jute and cotton fiber. When jute and cotton fiber were added with the pineapple fiber, the stress was transferred among pineapple, jute and cotton fiber instead of only pineapple fiber, thus, stress was reduced for matrix PP. Which is the cause of the change in mechanical properties. The better mechanical properties in this study came mainly from the sandwich structure, where the core layer (fabric) contributed a lot to the mechanical properties. From this investigation, it was clear that pineapple leaf, cotton and jute fiber based PP composites achieved more than double the mechanical property values over the matrix material alone. The content of cellulose in pineapple, jute and cotton fiber gave an advantage in this case [2].

Table 1. Tensile properties of polypropylene (PP) and fabric/PP composites.

Materials	Tensile Properties		
	Tensile Strength (MPa)	Tensile Modulus (MPa)	Elongation at Break (%)
PP	28 ± 1.5	338 ± 25	75 ± 4.75
Fabric/PP	58 ± 3.15	867 ± 56	22.38 ± 1.6

Table 2. Bending properties of polypropylene (PP) and fabric/PP composites.

Materials	Bending Properties	
	Bending Strength (MPa)	Bending Modulus (MPa)
PP	20 ± 1.45	230 ± 16
Fabric/PP	42 ± 2.78	495 ± 26

3.2. Water Absorption Profile of the Composites

The water absorption test allowed divulging the water absorbing behavior of the composite. The results of water absorption tests of the three samples of the fabric/PP composites are depicted in Figure 3, against the time of soaking in water at room temperature. With an increase of soaking time, the level of water absorption increased up to 40 min, as shown in the graph, after which no further water absorption occurred [45]. It was found that the fabric/PP composite absorbed 1.6% water. Motaleb et al. reported that PALF/PP composite (30% fiber content) absorbed no further water after 50 min and that the water uptake percentage was 1.46% [14]. Little variation was noticed for the fabric/pp composite due to its different fiber composition.

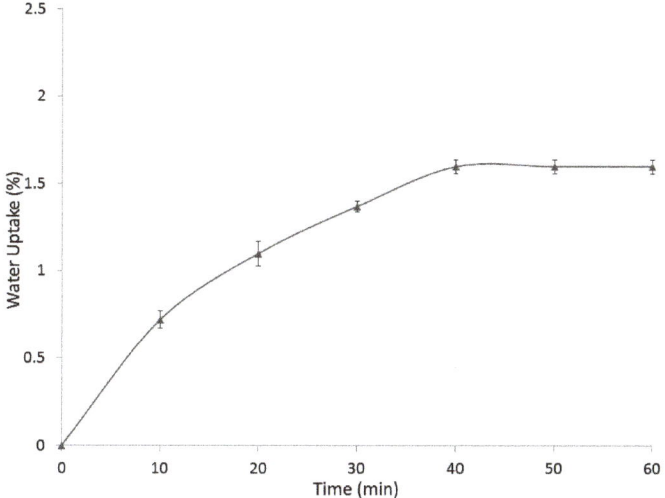

Figure 3. Water uptake % of fabric/PP composites.

Water absorption can be described by hydroxyl (–OH) groups that are present in the fiber cellulose resulting in the strong hydrophilic nature of fabric [46]. Polypropylene is strongly hydrophobic in nature [3] and resists water penetration when fabricating composites by sandwiching the fabric in between polypropylene sheets. Water was absorbed by the cut edge of the composite and this resulted in lower fabric/PP composite water uptake values.

3.3. Effect of Alkali

The effect of alkali (3%, 5% and 7% NaOH) on composites was tested at room temperature for 24 h. Tensile strength (TS), tensile modulus (TM), elongation at break (Eb%), bending strength (BS) and bending modulus (BM) values are shown in Figure 4. It was found that the mechanical properties of fabric/PP composites decreased for all conditions under the study. In this study, the mechanical properties of the composites decreased significantly when the composites were treated in 7% NaOH solution as shown in the figures. In particular, after 24 h of 7% alkali treatment, fabric/PP composite values decreased by 21%, 29%, 32%, 23% and 27% of TS, TM, Eb%, BS and BM respectively.

When natural fiber was alkali treated, which contains cellulose, crystal structure of fibers, the mechanical properties increased greatly. On the other hand, when the natural fiber reinforced composites were treated with alkali, there was some variation in the mechanical properties [47].

The change in mechanical properties can be well explained by the help of mercerization. It may be that the fabric lost its strength over time. Due to mercerization, the breaking tendency of composites was increased [2].

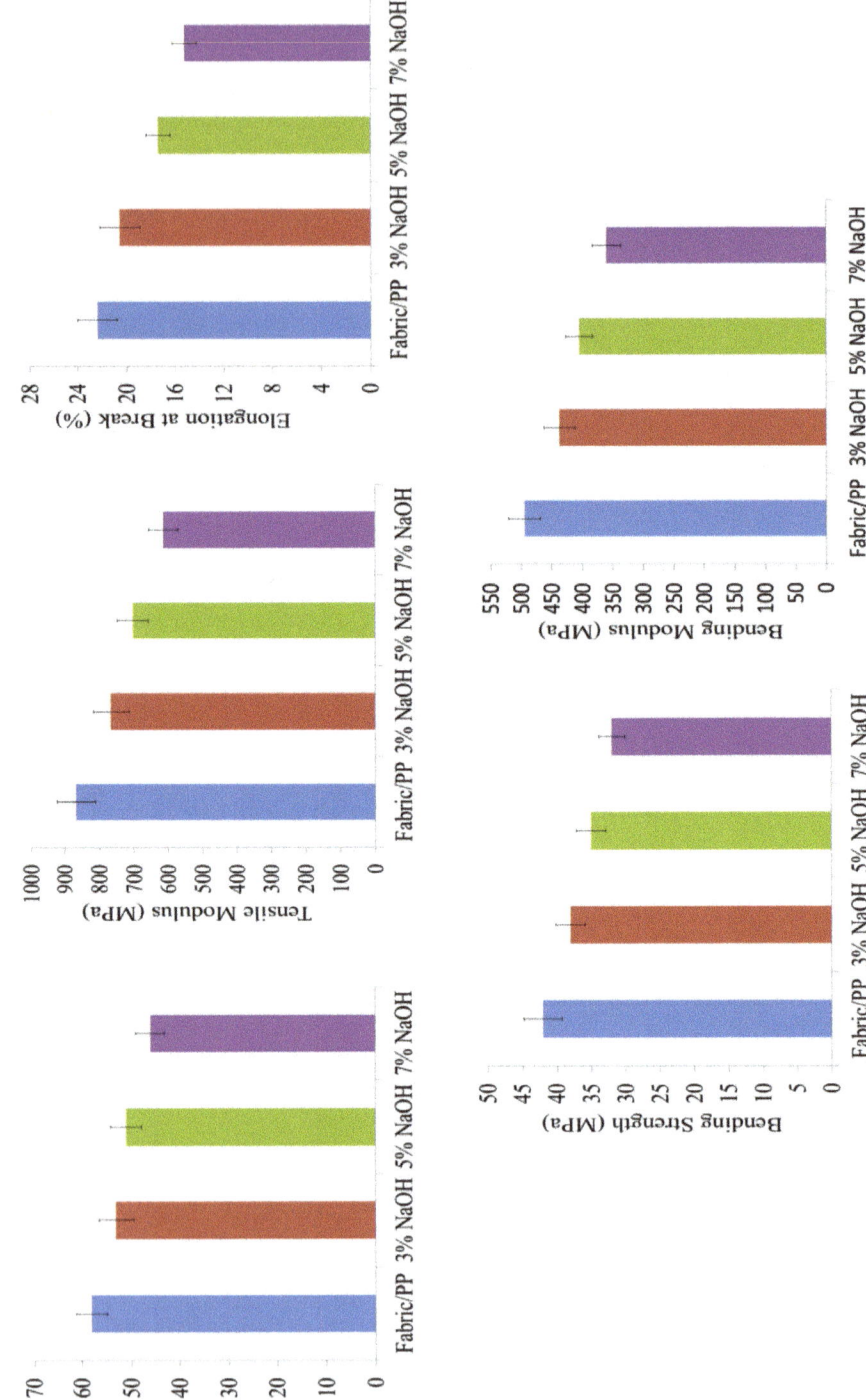

Figure 4. Effect of alkali on the mechanical properties of the composites.

3.4. Soil Degradation Test of the Composites

The degradation tests on the composites were carried out in soil at ambient conditions for up to 24 weeks. Tensile strength (TS), tensile modulus (TM) and elongation at break (Eb%) values are shown in Figure 5. From the figures, it can be seen that the TS, TM and Eb% of the fabric/PP composites decreased slowly with time. Fabric/PP composites consumed almost 50%, 48% and 49% of their TS, TM and Eb% respectively after 24 weeks of soil degradation. Similarly, decreased values of bending strength (BS) and bending modulus (BM) were also observed, and the results are depicted in Figure 5. Results revealed that soil degraded 45% and 47% of the initial BS and BM values of the fabric/PP composites respectively, after 24 weeks. It was also observed that, fabric/PP composites lost almost 41% of their mass after 24 weeks. For being a cellulose based natural biodegradable fiber, pineapple leaf, cotton and jute absorbs water within a few minutes, which indicates its strong hydrophilic nature. Cellulose has a strong aptitude to degrade in soil [48]. PP is strongly hydrophobic in nature. During the soil-degradation tests, the penetration of water occurred from the cut edges of the composites and degraded the cellulose resulting in significantly reduced mechanical properties. S. Nahar et al. stated that 40%, 46%, 36% and 35% of the TS, TM, BS and BM, respectively, were lost after a 24 week soil degradation test of natural fiber (jute fiber) reinforced pp based composite (50% fiber weight) [8]. Analyzing these values, no significant change of tensile properties are seen due to soil degradation test after 24 weeks. But some change of bending properties were happened because of different fiber percentage and composition of fabric/PP composite.

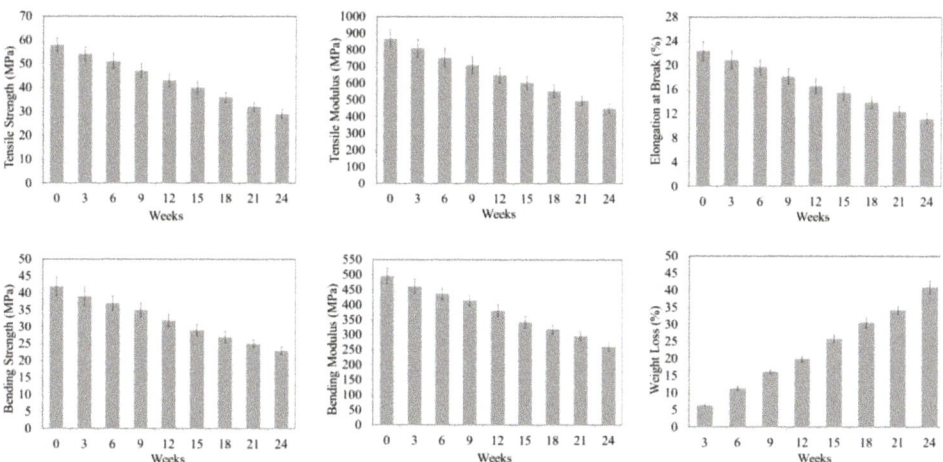

Figure 5. Soil degradation test of the composites.

4. Conclusions

Fabric-reinforced polypropylene-based composites were successfully prepared and characterized. Increased mechanical properties of the composites were seen compared with the matrix material. The TS and TM of the fabric/PP composites were found to be 58 MPa and 867 MPa respectively, compared with the PP matrix, a 107% and 156% increase of TS and TM was noticed. For the fabric/PP composites, BS and BM were found to be 42 MPa and 495 MPa respectively, which was 110% and 115% higher than those of the matrix material. Little variation was noticed in the water uptake, whereas reduced mechanical properties were observed, due to the addition of jute and cotton fiber compared with a 30 wt% PALF/PP composite [14]. Alkali reduced the mechanical properties of fabric/PP composites. Degradation testing of the fabric/PP composites were carried out for 6 months in soil and it was observed that composites retained about 50% of their original mechanical properties. Further research is needed in order to compare synthetic and natural fiber reinforced composites in order to improve knowledge on the topic.

Author Contributions: Conceptualization, M.B.H. and S.; methodology, M.B.H. and S.; validation, A.B.M.H.A. and H.M.; formal analysis, M.B.H. and A.N.; investigation, M.B.H. and A.N.; writing—original draft preparation, M.B.H.; writing—review and editing, M.B.H. and S.; supervision, S.

Funding: This research received no external funding.

Conflicts of Interest: The authors declare no conflicts of interest.

References

1. Sahadat Hossain, M.D.; Sarwaruddin Chowdhury, A.M.; Khan, R.A. Effect of disaccharide, gamma radiation and temperature on the physico-mechanical properties of jute fabrics reinforced unsaturated polyester resin-based composite. *Radiat. Eff. Defects Solids* **2017**, *172*, 517–530. [CrossRef]
2. Hoque, M.B.; Sahadat Hossain, M.D.; Nahid, A.M.; Solaiman, B.; Khan, A.R. Fabrication and characterization of pineapple fibre-reinforced polypropylene based composites. *Nano Hybrids Compos.* **2018**, *21*, 31–42. [CrossRef]
3. Sahadat Hossain, M.D.; Uddin, M.B.; Razzak, M.D.; Sarwaruddin Chowdhury, A.M.; Khan, R.A. Fabrication and characterization of jute fabrics reinforced polypropylene-based composites: Effects of ionizing radiation and disaccharide (sucrose). *Radiat. Eff. Defects Solids* **2017**, *172*, 904–914. [CrossRef]
4. Wu, L.-P.; Kang, W.-L.; Chen, Y.; Zhang, X.; Lin, X.-H.; Chen, L.-Y.; Gai, Y.-G. Structures and properties of low-shrinkage polypropylene composites. *J. Ap. Polym. Sci.* **2016**, *134*, 44275. [CrossRef]
5. Uawongsuwan, P.; Yang, Y.; Hamada, H. Long jute fiber-reinforced polypropylene composite: Effects of jute fiber bundle and glass fiber hybridization. *J. App. Polym. Sci.* **2014**, *132*, 41819. [CrossRef]
6. Varshney, D.; Debnath, K.; Singh, I. Mechanical characterization of polypropylene (PP) and polyethylene (PE) based natural fiber reinforced composites. *Int. J. Surf. Eng.* **2014**, *4*, 16–23.
7. Nuruzzaman Khan, M.D.; Roy, J.K.; Akter, N.; Zaman, H.U.; Islam, T.; Khan, R.A. Production and properties of short jute and short e-glass fiber reinforced polypropylene-based composites. *Open J. Compos. Mater.* **2012**, *2*, 40–47. [CrossRef]
8. Nahar, S.; Khan, R.A.; Dey, K.; Sarker, B.; Das, A.K.; Ghoshal, S. Comparative studies of mechanical and interfacial properties between jute and bamboo fiber- Reinforced polypropylene-based composites. *J. Thermoplast. Compos. Mater.* **2012**, *25*, 15–32. [CrossRef]
9. Khan, M.A.; Hinrichsen, G.; Drzal, L.T. Influence of noble coupling agents on mechanical properties of jute reinforced polypropylene composites. *J. Mater. Sci. Lett.* **2001**, *20*, 1711–1713. [CrossRef]
10. Kasim, A.N.; Selamat, M.Z.; Daud, M.A.M.; Yaakob, M.Y.; Putra, A.; Sivakumar, D. Mechanical properties of polypropylene composites reinforced with alkaline treated pineapple leaf fibre from Josapine cultivar. *Int. J. Automotive Mech. Eng. IJAME* **2016**, *1*, 3157–3167. [CrossRef]
11. Kasim, A.N.; Selamat, M.Z.; Aznan, N.; Sahadan, S.N.; Daud, M.A.M.; Jumaidin, R.; Salleh, S. Effect of pineapple leaf fiber loading on the mechanical properties of pineapple leaf fiber–polypropylene composite. *J. Teknologi* **2015**, *77*, 117–123. [CrossRef]
12. Ranganathan, N.; Oksman, K.; Nayak, S.K.; Sain, M. Regenerated cellulose fibers as impact modifier in long jute fiber reinforced polypropylene composites: Effect on mechanical properties, morphology, and fiber breakage. *J. App. Polym. Sci.* **2014**, *132*, 41301. [CrossRef]
13. Berhanu, T.; Kumar, P.; Singh, I. Mechanical behaviour of jute fibre reinforced polypropylene composites. In Proceedings of the 5th International and 26th All India Manufacturing Technology, Design and Research Conference (AIMTDR 2014), IIT Guwahati, Assam, India, 12–14 December 2014.
14. Motaleb, K.Z.M.A.; Islam, S.; Hoque, M.B. Improvement of physicomechanical properties of pineapple leaf fiber reinforced composite. *Int. J. Biomater.* **2018**. [CrossRef] [PubMed]
15. Soutis, C. Fibre reinforced composites in aircraft construction. *Prog. Aerosp. Sci.* **2005**, *41*, 143–151. [CrossRef]
16. Lau, K.-T.; Hung, P.-Y.; Zhu, M.H.; Hui, D. Properties of natural fibre composites for structural engineering applications. *Compos. Part B Eng.* **2018**, *136*, 222–233. [CrossRef]
17. Sfondrini, M.F.; Cacciafesta, V.; Scribante, A. Shear bond strength of fibre-reinforced composite nets using two different adhesive systems. *Eur. J. Orthod.* **2011**, *33*, 66–70. [CrossRef]

18. Cacciafesta, V.; Sfondrini, M.F.; Lena, A.; Scribante, A.; Vallittu, P.K.; Lassila, L.V. Flexural strengths of fiber-reinforced composites polymerized with conventional light-curing and additional postcuring. *Am. J. Orthod. Dentofac. Orthop.* **2007**, *132*, 524–527. [CrossRef]
19. Lei, W.; Fang, C.; Zhou, X.; Li, Y.; Pu, M. Polyurethane elastomer composites reinforced with waste natural cellulosic fibers from office paper in thermal properties. *Carbohydr. Polym.* **2018**, *197*, 385–394. [CrossRef]
20. Anand, R.S.; Daniel, F.C.; Rodney, E.J.; Roger, M.R. Renewable agricultural fibers as reinforcing fillers in plastics: Mechanical properties of kenaf fibers–polypropylene composites. *Ind. Eng. Chem. Res.* **1995**, *34*, 1889–1896.
21. Ma, C.M.; Tseng, H.; Wu, H. Blocked diisocyanate polyester-toughened novolak-type phenolic resin: Synthesis, characterization, and properties of composites. *J. Appl. Polym. Sci.* **1998**, *69*, 1119–1127. [CrossRef]
22. Valadez-Gonzalez, A.; Cervantes-Uc, J.M.; Olayo, R.; Herrera-Franco, P.J. Chemical modification of henequen fibers with an organosilane coupling agent. *Compos. Part B* **1999**, *30*, 321–331. [CrossRef]
23. Mohanty, A.K.; Misra, M.; Drzal, L.T. *Natural Fibers, Biopolymers and Biocomposites*; Taylor & Francis, CRC Press: Boca Raton, FL, USA, 2005.
24. Sreekala, M.S.; Kumaran, M.G.; Thomas, S. Oil palm fibers: Morphology, chemical composition, surface modification, and mechanical properties. *J. Appl. Polym. Sci.* **1997**, *66*, 821–835. [CrossRef]
25. Herrera-Franco, P.J.; Valadez-Gonzalez, A. A study of the mechanical properties of short natural-fiber reinforced composites. *Compos. Part B Eng.* **2005**, *36*, 597–608. [CrossRef]
26. Abdelmouleh, M.; Boufi, S.; Belgacem, M.N.; Dufresne, A. Short natural-fibre reinforced polyethylene and natural rubber composites: Effect of silane coupling agents and fibres loading. *Compos. Sci. Technol.* **2007**, *67*, 1627–1639. [CrossRef]
27. Tserki, V.; Zafeiropoulos, N.E.; Simon, F.; Panayiotou, C. A study of the effect of acetylation and propionylation surface treatments on natural fibres. *Compos. Part A Appl. Sci. Manuf.* **2005**, *36*, 1110–1118. [CrossRef]
28. Yan, L.; Chouw, N.; Yuan, X. Improving the mechanical properties of natural fibre fabric reinforced epoxy composites by alkali treatment. *J. Reinf. Plast. Compos.* **2012**, *31*, 425–437. [CrossRef]
29. Liu, Q.; Stuart, T.; Hughes, M.; Sharma, H.S.S.; Lyons, G. Structural biocomposites from flax-part II: The use of PEG and PVA as interfacial compatibilising agents. *Compos. Part A Appl. Sci. Manuf.* **2017**, *38*, 1403–1413. [CrossRef]
30. Carus, L.S.M. Targets for bio-based composites and natural fibres. *JEC Compos. Mag.* **2011**, *8*, 31.
31. Mohanty, K.; Misra, M.; Drzal, L.T. *Natural Fibres, Biopolymers and Biocomposites*; Taylor & Francis, CRC Press: Oxfordshire, UK, 2005.
32. Satyanarayana, K.G.; Pillai, S.K.G.; Pai, B.C.; Sukumaran, K. Lignocellulosic fibre reinforced polymer composite. In *Handbook of Ceramic and Composites*; Cheremisinoff, N.P., Ed.; Marcel Dekker: New York, NY, USA, 1990.
33. Satyanarayana, K.G.; Sukumaran, K.; Mukherjee, P.S.; Pavithran, C.; Pillai, S.K.G. Natural fibre-polymer composites. *Cem. Concr. Compos.* **1990**, *12*, 117–136. [CrossRef]
34. Arib, R.M.N.; Sapuan, S.M.; Hamdan, M.A.M.M.; Paridah, M.T.; Zaman, H.M.D.K. A literature review of pineapple fibre reinforced polymer composites. *Polym. Polym. Compos.* **2004**, *12*, 341–348. [CrossRef]
35. Pavithran, C.; Mukherjee, P.S.; Brahmakumar, M.; Damodaran, A.D. Impact properties of natural fibre composites. *J. Mater. Sci. Lett.* **1987**, *6*, 882–884. [CrossRef]
36. Mishra, S.; Misra, M.; Tripathy, S.S.; Nayak, S.K.; Mohanty, A.K. Potentiality of pineapple leaf fibre as reinforcement in PALF-polyester composite: Surface modification and mechanical performance. *J. Reinf. Plast. Compos.* **2001**, *20*, 321–334. [CrossRef]
37. Harzallah, O.; Benzina, H.; Drean, J.-Y. Physical and mechanical properties of cotton fibers: Single-fiber failure. **2009**, *80*, 1093–1013. [CrossRef]
38. Azwa, Z.N.; Yousif, B.F.; Manalo, A.C.; Karunasena, W. A review on the degradability of polymeric composites based on natural fibres. *Mater. Des.* **2013**, *47*, 424–442. [CrossRef]
39. Khan, R.A.; Khan, M.A.; Sultana, S.; Noor, F.G. Mechanical, degradation, and interfacial properties of synthetic degradable fiber reinforced polypropylene composites. *J. Reinf. Plast. Compos.* **2009**, *29*, 466–476. [CrossRef]
40. Garcia, M.; Vliet, G.V.; Jain, S.; Zyl, W.E.V.; Boukamp, B. Polypropylene/SiO_2 nano composites with improved mechanical properties. *Rev. Adv. Mater. Sci.* **2004**, *6*, 169–175.

41. Karmaker, A.C.; Hinrichsen, G. Processing and characterization of jute fiber reinforced thermoplastic polymers. *Polym. Plast. Technol. Eng.* **1999**, *30*, 609–621. [CrossRef]
42. Bledzki, A.K.; Gassan, J. Composites reinforced with cellulose based fibers. *J. Prog. Polym. Sci.* **1999**, *24*, 221–274. [CrossRef]
43. Wambua, P.; Ivan, J.; Verport, I. Natural fibers: Can they replace glass in fiber reinforced plastics. *J. Compos. Sci. Technol.* **2003**, *63*, 1259–1264. [CrossRef]
44. Czvikovszky, T. Reactive recycling of multiphase polymer systems. *Nucl. Instrum. Methods Phys. Res. B* **1995**, *105*, 233–237. [CrossRef]
45. Zhu, J.; Abhyankar, H.; Njuguna, J. Effect of fibre treatment on water absorption and tensile properties of flax/tannin composites. In Proceedings of the 11th International Conference on Manufacturing Research (ICMR2013), Bedford, UK, 19–20 September 2013; pp. 387–392.
46. Kumar, B.; Lin, S.T. Redox state of iron and its related effects in the $CaOP_2O_5$-Fe_2O_3 glasses. *J. Am. Ceram. Soc.* **1991**, *74*, 226–229. [CrossRef]
47. Ray, D.; Sarker, B.K.; Rana, A.K.; Bose, N.R. Effect of alkali treated jute fibres on composite properties. *Bull. Mater. Sci.* **2001**, *24*, 129–135. [CrossRef]
48. Khan, R.A.; Parsons, A.J.; Jones, I.A.; Walker, G.S.; Rudd, C.D. Surface teatment of phosphate glass fibers using 2-Hydroxyethyl methacrylate: Fabrication of poly(caprolactone)-based composites. *J. Appl. Polym. Sci.* **2009**, *111*, 246–254. [CrossRef]

© 2018 by the authors. Licensee MDPI, Basel, Switzerland. This article is an open access article distributed under the terms and conditions of the Creative Commons Attribution (CC BY) license (http://creativecommons.org/licenses/by/4.0/).

Review

Jute Based Bio and Hybrid Composites and Their Applications

Muhammad Ahsan Ashraf [1], Mohammed Zwawi [2], Muhammad Taqi Mehran [1], Ramesh Kanthasamy [3] and Ali Bahadar [3,*]

1. School of Chemical and Materials Engineering (SCME), National University of Sciences and Technology (NUST), Islamabad 44000, Pakistan
2. Department of Mechanical Engineering, King Abdulaziz University, Rabigh 21911, Saudi Arabia
3. Department of Chemical and Materials Engineering, King Abdulaziz University, Rabigh 21911, Saudi Arabia
* Correspondence: absali@kau.edu.sa or engrbahadur@gmail.com

Received: 19 June 2019; Accepted: 22 August 2019; Published: 28 August 2019

Abstract: The popularity of jute-based bio and hybrid composites is mainly due to an increase in environmental concerns and pollution. Jute fibers have low cost, high abundance, and reasonable mechanical properties. Research in all-natural fibers and composites have increased exponentially due to the environment concerns of the hazards of synthetic fibers-based composites. Jute based bio and hybrid composites have been extensively used in number of applications. Hybrid jute-based composites have enhanced mechanical and physical properties, reasonably better than jute fiber composites. A detailed analysis of jute-based bio and hybrid composites was carried out in this review. The primary aim of this review paper is to provide a critical analysis and to discuss all recent developments in jute-based composites. The content covers different aspects of jute-based composites, including their mechanical and physical properties, structure, morphology, chemical composition, fiber modification techniques, surface treatments, jute based hybrid composites, limitations, and applications. Jute-based composites are currently being used in a vast number of applications such as in textiles, construction, cosmetics, medical, packaging, automobile, and furniture industries.

Keywords: hybrid bio-composite; natural fiber; extrusion; jute; renewable

1. Introduction

We are surrounded by environmental pollution, which has now become a major threat to all living creatures. Leading nations are trying to minimize pollution by taking radical steps to replace pollution-causing materials with renewable ones. Researchers have made it possible to replace conventional synthetic materials with natural bio-based alternatives. Composite materials are considered to be one of the most important materials in diversified and load-bearing applications. Many resources have been spent on developing synthetic composites, which have performed well in different applications. But now, with ever-increasing environmental concerns and threats, much focus has been diverted to the development of bio- and hybrid composites. As a result, in the past few years, a great deal of attention has been paid to the development of these composites. Currently, bio-composites are being developed to meet the performance of synthetic composites. These bio-composites are biodegradable and various natural fibers have been used for their fabrication, aiming at making them eco-friendly in nature and with minimum carbon emissions. Even hybrid composites have also become popular due to superior mechanical and physical properties compared to pure bio-composites. Bio-composites are natural, light in weight, low in carbon emissions, and low in material and manufacturing costs. And above all, these fibers used in the manufacturing of bio-composites are abundant in nature. Bio-composites are on the verge of becoming an integral part of society due to their various useful applications [1]. Consequently, bio-composites have emerged as the best replacement for synthetic composites. Different bio-composites are at various developmental stages, both in research and in the industrial community. Many bio-composite materials have shown mechanical and physical properties comparable to that of synthetic composites. These natural fibers are adequate replacements for glass and carbon fibers. Despite swift progress in this field, many shortcomings and limitations still exist, hindering the practical implementation of these natural fibers. The limitations of bio-based polymers and composites include poor adhesion between polymer and matrix, incompatibility of fibers, agglomeration of fibers, and a lack of manufacturing processes [2,3]. Research is being carried out in developed and developing countries with the aim of overcoming the outstanding issues related to bio-composites. Many countries are taking further steps ahead by encouraging industries to use bio-composites and also by giving them different trading incentives, while industries themselves are adopting these bio-composites for various international certifications and for gaining access to international markets. Biodegradable materials will not only benefit us environmentally but will also help to improve economic outlooks [4].

Recently, bio-composites and bio-degradable materials have gained much momentum. These materials are subjected to continuous research development regarding numerous industrial sectors. Different plant fibers are shown in Figure 1 [5–8]. Among various different natural fibers, jute is one of the most important fibers used in the manufacturing of bio-composites. Jute belongs to the bast fiber family and is normally grown in the tropical areas of China, Bangladesh, India, and Indonesia. It is considered to be one of the most produced fibers in the world [9]. Jute can easily be grown in a humid atmosphere and warm temperature range. Additionally, jute is known as a rainy season crop and can survive flood conditions. Jute is solely grown for the extraction of fibers [6]. Jute plants are differentiated based on plant type, color, strength, and length of fibers. White and Tossa jute are the two main types of jute plants [10] which are cultivated in tropical and high-temperature areas. Asia is the biggest producer of jute and contributes to around 95% of total jute production in the world [6]. Jute fibers have gained much attention in the last few years due to their physical and mechanical properties. The mechanical properties of jute fibers are believed to be comparable to glass fiber in terms of specific strength and specific modulus [11].

In addition, jute fibers are eco-friendly, renewable, cheap to produce, and bio-degradable [12]. The main constituents of jute fibers are cellulose and lignin. Cellulose is a polysaccharide that helps to form hydrogen bonding between matrix and natural-fiber-improving interfacial adhesions [13]. Jute fiber has a high demand from textiles other than the composite and bio-polymer industries [14]. The properties of jute fibers are usually determined by the maturity of the plant, fiber length, and the processing techniques used for the manufacturing of composites [6]. Jute fiber is currently being used in many applications including textiles, automobiles, and even in some load-bearing applications. In the automobile sector, bio-composites and bio-polymers of jute are used to produce different components such as door panels, trunk liners, and cup holders [15,16]. Even big auto manufacturers such as Mercedes, along with many European and American car manufacturers, are keen to use more renewable composites and polymers [17]. The annual production of various natural fibers is shown in Table 1 [6,18–20].

Over the last few years, jute-based hybrid composites containing natural components have emerged with better mechanical properties than natural jute-based composites. Usually hybrid composites are made from a combination of natural and synthetic fibers in a matrix while bio-composites consist of natural fibers and synthetic or natural material composite. Hybrid composites contain less percentage of natural fibers but the shortcomings of all-natural fiber composites can be overcome by reducing natural fiber content with the addition of synthetic fibers such as glass fibers.

Table 1. Annual production of different fibers.

Fiber Source	Fiber Type	Annual Production (10^3 Tonnes)
Jute	Bast and Core fiber	3600
Bamboo	Wood fiber	30,000
Sugar cane	Wood or Stem fiber	75,000
Grass	Grass fiber	700–750
Ramie	Bast fiber	100–110
Abaca	Leaf fiber	70–90
Hemp	Bast and Core fiber	200–220
Sisal	Leaf fiber	370–380
Coir	Seed fiber	600–650
Kenaf	Bast and Core fiber	950–990
Flax	Bast fiber	830

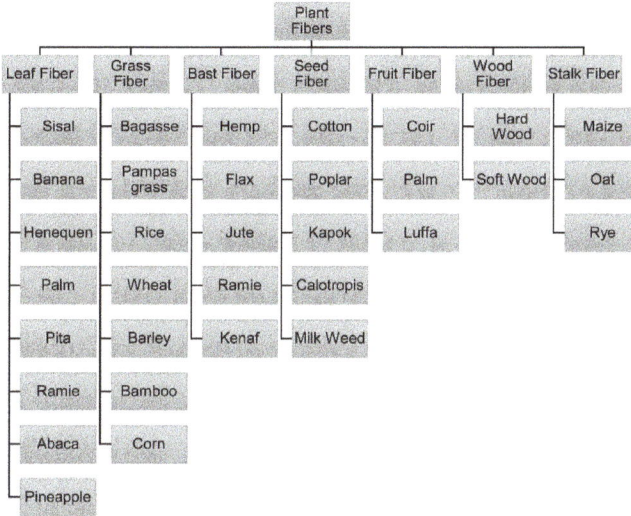

Figure 1. Different types of plant fibers.

In this review, we aim to discuss recent developments regarding jute-based bio and hybrid composites. The main objective of this review is to provide a critical assessment of the literature and to summarize the key findings of different jute-based research studies. This review paper contains the mechanical and physical properties, structure, morphology and chemical composition, fiber-modification techniques, surface treatments, limitations and applications of jute fiber-based bio- and hybrid composites. The present work also focuses on methods to improve properties by overcoming the limitations associated with these jute-based composites. Due to huge number of potential applications, jute-based composites have become a topic of interest in the research community. The applications of these composites include textiles, automobiles, polymers, medical, cosmetic, and construction industries.

2. Structure, Morphology, and Chemical Composition of Jute

Jute fibers, obtained from jute plants, are commonly known as lignocellulosic fibers [21]. Jute fibers are multi-cellular and can be found in the bast region of jute plants, stretching along the stem [22,23]. A single jute fiber is formed by combining different cells and constituents like lignin, cellulose, oils, waxes, and the different types of fats [24]. These cementing constituents help in the formation of a network of jute fibers in the stem region and this network of jute fibers is called a strand. Jute fibers join with one another to form long fibers and a meshy network [25]. The geometry and composition of jute fibers can vary from plant to plant due to differences in growth conditions. The composition can even vary from fiber to fiber in a single plant [25]. Jute fiber structure, length, and chemical composition are dependent on factors like growth environment, weather conditions, defects, plant maturity, extraction, and the modification methods used [2]. Lumen is a central hollow cavity present in a jute fiber cell, which produces fiber with a low density [26,27]. Each unit cell of fiber is connected with other cells through the middle lamella, which is usually composed of lignin and cellulose. This middle lamella is responsible for joining different unit cells [28].

Jute fiber is composed of different types of polysaccharides and lignin. Polysaccharides are made up of hemicellulose and alpha cellulose [29]. In jute fiber, cellulose is the primary constituent of the cell wall and consists of different glucose rings. Glucose rings form cellobiose consisting of repeating units of glucose dimers [30]. Linear polymerization of glucose rings forms a cell wall and ultimately the jute fiber. These glucose rings interact with one another through covalent bonds that are responsible for different physical and mechanical properties [31]. Hydroxyl groups present in this glucose and the cellulose chain interacts with other hydroxyl groups and water molecules, forming hydrogen bonds. Hydrogen bonds between these hydroxyl groups are also responsible for the hydrophilic nature, crystallization, and three-dimensional structure of jute fiber. Cellulose has a hydrophilic nature and is dissolvable in water. Due to these factors, water is absorbed in cellulose, which can cause an overall swelling of the fiber [32,33]. Jute fiber is acidic in nature due to the presence of hemicellulose and poly-uronic acid [34]. The close and microscopic examination of unit cells revealed that crystallization is not homogenous throughout the fiber. Highly-ordered and closely-packed regions are known as micro-fibril structures, while loosely packed and non-homogenous regions are usually called fringed fibril areas. Highly-ordered regions are crystalline while non-homogenous regions are amorphous. A micro-fibril angle with respect to plant length is also responsible for different mechanical and physical properties [35]. Cellulose can be divided into cellulose I alpha, cellulose I beta, cellulose II, and cellulose III. The cell wall consists of a primary and secondary cell wall. The primary cell wall is usually thin while the secondary cell wall is usually a thick layer, though both are formed by a combination of very fine micro-fibrils [36]. The primary cell wall has a crisscross-linking of micro-fibrils while the secondary cell wall has highly-ordered micro-fibril arrangements [37,38]. Microfibrils have nano-crystals of cellulose I alpha.

Figure 2 shows the cross-section of the jute fiber.

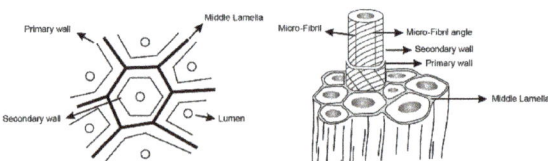

Figure 2. Showing cross section of jute fiber. Reproduced with permission from Roy and Lutfar [39].

Jute fiber is mainly composed of lignin, cellulose, waxes, pectin, protein, nitrogenous compounds, and mineral and inorganic matter. Cellulose is further divided into alpha cellulose and hemicellulose, with these constituents varying slightly from plant to plant and in different growing conditions. Cellulose is the main constituent of jute fiber. Normally, a jute fiber contains around 59–63% of alpha cellulose, 22–26% of hemicellulose, 12–14% lignin, 0.4–0.8% of waxes, 0.2–0.5% of pectin, 0.6–1.2% of mineral matter and traces of other constituents [18,40,41]. Glucose is a basic building unit of cellulose, which cannot be easily oxidized though acids can hydrolyze the substance. Alpha cellulose consists of long cellulose chains with a combination of high molecular weight polysaccharides [4], while hemicellulose is formed by the combination of a small-length cellulose chain with low molecular weight polysaccharides. Hemicellulose has a branched structure, giving rise to an amorphous nature. Hemicellulose is believed to act as a binder between different constituents in jute fiber, such as micro-fibrils and lignin, while pectin gives plant structure its elasticity. Jute fiber also contains some minor traces of waxes, which are insoluble in water and acids [14,42,43].

Lignin is known to be one of the most abundant polymers [44], which is considered to be a complex three-dimensional structure with a presence of aliphatic and aromatic groups. Lignin consists of a high amount of carbon along with a low amount of hydrogen [45]. These amounts of carbon and hydrogen make lignin aromatic or unsaturated in nature with the presence of hydroxyl, carboxyl, and methoxyl groups. The basic unit cell of lignin is composed of hydroxyl and methoxyl groups, which make the constituent complex and amorphous. It is readily soluble in alkali and can be oxidized and condensed, though it cannot be hydrolyzed by an acidic medium [4]. Lignin is connected with other constituents of jute fiber through being alkali resistant and having alkali-sensitive links. Alkali-sensitive linkage is in-between the hydroxyl groups of lignin and the carboxyl groups of cellulose or hemicellulose, while the alkali-resistant link is actually a crosslinking of the hydroxyl groups of lignin and cellulosic constituents [46]. Lignin usually provides support to plants and has a hydrophobic nature.

3. Mechanical Properties of Jute Fiber

Jute fiber is considered to be one of the most important fibers for the production of bio-composites and bio-plastics. Much research can be found studying the different mechanical properties of jute fiber, which have acceptable mechanical properties like tensile properties, specific strength, and modulus, hence increasing its potential use in different applications [47]. The values of the different mechanical properties of jute fiber reported by researchers are listed below in Table 2. The addition of synthetic fibers in jute-based composites are found to increase its mechanical properties. These hybrid composites have shown better results and mechanical properties than that of bio-composites made up of jute fiber [36]. Hybrid composites are the combination of two or more natural and synthetic fibers with matrix material [48]. Synthetic fiber helps to balance the shortcomings associated with natural fibers [49] and helps to increase the mechanical properties and to decrease costs associated with composites [4,26]. Mechanical properties are dependent on many factors, one of the most important factors is the fiber length of composite [50]. The critical length of fiber is important to have better mechanical properties, stress transfer, and good fiber/matrix. Fiber length is critical for carrying maximum load. Fiber length beyond critical length results in poor fiber/matrix adhesion and poor stress transfer which will result in failure and the pre-mature fracture of fibers [51]. Chollakup et al. [52] studied the effects of long and short fibers of pineapple leaf on properties of composites. Composites with longer fiber length were found to be stronger in comparison with short fibers. Longer fibers exhibited homogenous dispersion while short fibers were heterogeneously dispersed. The tip or end of short fibers behaves as a stress concentrating site, leading to poor stress and load transfer from matrix to fibers. Controlling fiber orientation and aggregation are some of the issues associated with short fibers [53]. Mishra et al. [54] studied a hybrid composite of jute with epoxy and found an increase in flexural, tensile, and impact strength. The composite was found to have better mechanical interlocking between fiber and matrix. Abdullah Al et al. [55] studied the mechanical properties of jute fiber with epoxy glass fiber. They found that the mechanical properties were improved with the addition of glass fiber. An ultraviolet radiation technique was used to further improve the mechanical properties of the jute/glass fiber composite. Ahmed and Vijayarangan [56] studied the different mechanical properties of the jute/glass fiber composite. Significant improvements were observed in mechanical properties with the addition of glass fiber. Ahmed et al. [57] studied the effects of glass fiber when added to jute fiber to form a hybrid composite. In this research, it was presented that mechanical properties like tensile, interlaminar shear, and flexural strength showed notable improvement with the addition of glass fiber. The addition of just 16.5 wt % of glass fiber improved the shear, tensile, and flexural strength properties by 17.6%, 37%, and 31.23%, respectively. This hybrid composite showed a better resistance to moisture. As jute fiber is hydrophilic in nature, which is a major hindrance to achieving better mechanical properties. The low mechanical properties of jute fiber are due to poor fiber–matrix adhesion, fire resistance, and thermal degradation [58]. Zamri et al. [59] studied the effects of water absorption on the mechanical properties of a jute/glass fiber composite. Mechanical properties like flexural and compression strengths were significantly declined after water absorption. Many surface treatment methods are reported in literature to improve the mechanical properties of jute fibers. Table 3 depicts the mechanical properties and surface treatments of jute-based composites. These methods include physical, chemical and physio-chemical, and mechanical surface modification methods. These methods are found to be effective in improving the mechanical properties of fibers [2]. Figure 3 shows a comparison of tensile strength of treated and untreated jute fiber composites. A jute/polylactide [60] composite was fabricated using 50 wt % jute fiber and treated with 5% aqueous NaOH solution. Alkali treatment increased roughness which improved fiber/matrix adhesion. A jute/polypropylene [61] composite was fabricated using 25 wt % jute and was post-treated with urea. Tensile strength increased significantly after the urea post treatment. Jute/epoxy [62] was fabricated by the hand lay-up technique and treated with 20% NaOH. A jute/polypropylene composite was fabricated using a hand–lay-up technique. Jute fibers were treated with 7% NaOH.

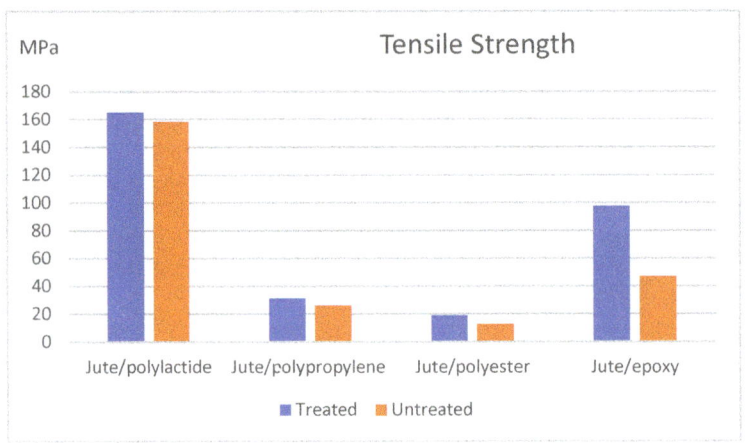

Figure 3. Showing tensile strength comparison of treated and untreated jute composite.

Table 2. Mechanical properties of jute fiber.

Mechanical Properties of Jute Fiber									
Density (g/cm^3)	Diameter (μm)	Micro-Fibrillar Angle (°)	Moisture Content (%)	Tensile Strength (MPa)	Tensile Modulus (GPa)	Specific Strength (MPa/g·m^{-3}) (S/ρ)	Specific Modulus (GPa/g·m^{-3}) (E/ρ)	Elongation at Break (%)	References
1.3–1.5	-	-	-	200–770	20–55	310–625	2–37	-	[33,63–66]
1.3–1.45	-	-	-	393–780	13–30	-	-	1.9	[67,68]
1.3–1.45	20–200	-	-	393–773	13–26.5	-	-	7–8	[69]
-	-	-	-	320–800	8–78	-	-	-	[36,40]
1.3	-	-	-	393–773	26.5	-	-	1.5–1.8	[70]
-	25–30	-	-	400–800	10–30	-	-	1.5–1.8	[2]
-	25–30	7–9	-	393–800	13–27	-	-	0.7	[71,72]
1.3–1.5	-	8	-	393–800	13–26.5	-	-	1.2–1.8	[73]
1.3–1.49	20–200	8	15.5–13.7	320–800	8–78	-	30	1–1.8	[74]
1.23	5–25	-	12	187–773	20–55	140–320	14–39	1.5–3.1	[2,46,75,76]
1.3–1.5	-	-	12	393–800	10–55	300–610	7.1–39	1.5–1.8	[33,77,78]

Table 3. Mechanical properties and surface treatments of different jute-based composites.

Jute Composite	Fabrication Method	Key Parameters and Findings	Mechanical Properties	Surface Treatments and Effects on Mechanical Properties	References
Jute/epoxy	Hand lay-up	Void content decreased and mechanical properties increased with the increase in jute fiber content in composite.	Properties like hardness, impact strength and tensile strength increased with the increase in jute fiber content due to improved fiber/matrix adhesion with better interlocking. Inclusion of fibers increases modulus of composite, increasing overall hardness.	-	[47]
Glass/jute fiber reinforced epoxy composite	Hand lay-up	Flexural load separated fibers from matrix upon failure.	Jute–glass fiber composite had tensile strength up to 63 MPa and flexural load up to 1.03 KN.	-	[68]
Jute fiber reinforced with epoxy and polyester matrices	Compression molding	20 wt % jute fibers and 80 wt % matrix materials were used with fiber length of 2–3 mm.	Jute epoxy composite had higher tensile strength, flexural strength and tensile modulus due to better stress distribution and fiber/matrix adhesion.	5% NaOH (alkali treated) jute fiber composites showed better tensile strengths and flexural strengths than 10% NaOH treated composites.	[36]
Jute/glass fiber reinforced epoxy composite	Hand lay-up	64–69% of epoxy resin, 18–31% of jute fibers, 0–19% of glass fibers were used in fabrication of different samples.	Increase in glass fiber content increased mechanical properties of composite. Composite with 64% epoxy resin, 18% of jute fibers and 19% of glass fiber had better tensile strength while flexural strength had not any significant change. Greater jute content led to rapid mass loss and increased moisture absorption.	-	[79]
Polylactide and jute composite	Solvent casting method	Samples with 50 wt % of jute were fabricated.	Tensile strength, tensile modulus, Izod impact strength, flexural strength and modulus for untreated jute/polylactide were 158 MPa, 5.3 GPa, 60 KJ/m^2, 180 MPa and 10.5 GPa respectively. Untreated samples had greater Izod impact strength and thermal stability due to fiber pull out mechanism. While treated samples had fractured fibers which require less energy than pull out mechanism.	Samples were modified by alkali, silane, peroxide and permanganate surface treatments. Silane treated samples had elevated values for tensile strength and modulus, flexural strength and modulus. Surface treatments increase surface roughness which in turn increases fiber/matrix adhesion and inter-locking. Surface treatment alters fiber structure and can chemically modify mechanical properties, increasing mechanical properties of composite.	[60]
Jute fiber/polypropylene composite	Extrusion	30 wt % of fiber loading had optimum mechanical properties.	Tensile strength showed decreasing trend as increase in jute fiber content increased the area of fiber/matrix interface, while tensile modulus increased due to obstruction in stress propagation by micro spaces. Flexural strength and modulus increased up to 30wt % fiber loading. Impact strength showed an increasing trend as larger force is required to pull out fibers up till 30 wt % of jute contents.	Urea treatment improved properties such as fiber/matrix adhesion, tensile strength and modulus, flexural strength and modulus and impact strength	[61]

Table 3. *Cont.*

Jute Composite	Fabrication Method	Key Parameters and Findings	Mechanical Properties	Surface Treatments and Effects on Mechanical Properties	References
Jute reinforced epoxy composite	Hand lay-up	Different mechanical and water absorption properties were studied for treated and untreated samples.	Tensile strength and flexural strength of untreated jute fiber epoxy composite were 46.7 and 62.4 MPa respectively.	Alkali treatment increased tensile strength by 108% and flexural strength by 28%.	[62]
Jute/glass epoxy composite	Hand lay-up	Jute and glass fibers were used in epoxy matrix to fabricate composite.	Hardness values increased with the increase in glass fiber content. Glass fibers have high hardness values which increase hardness of composite. Tensile strength showed increasing trend with the increase in glass fiber composition. Natural fibers enhanced degradability properties, and glass fibers enhanced brittleness.	Strength of composite increased by 11% when fibers were treated with NaOH. Tensile strength for treated composite increased due to removal of fiber components such as hemicellulose, lignin along with amorphous and crystalline parts of fibers.	[80]
Jute/Epoxy glass composite	Hand lay-up	Homogenous thickness of samples was obtained through compression technique.	Jute/E-glass composite showed better tensile strength than pure jute-based composite due to better stress transfer. Jute fibers increased toughness and decreased brittleness. While epoxy glass improved erosion wear properties.	Jute fibers were alkali treated to get rid of lignin, hemicellulose, and cellulose from the surface of fibers.	[81]
Jute/kenaf fibers reinforced epoxy composite	Hand lay-up	Samples were prepared using 56% of jute and kenaf fibers, 40% of epoxy, and 4% of hardeners.	Flexural strength, impact strength, tensile and compressive strength of treated fibers had enhanced values as compared untreated fibers composite.	Surface treatment of kenaf and jute fibers removed hemi-cellulose, pectin and other non-cellulosic matter. Surface treatment increased surface area, reduced moisture absorption and improved roughness of fibers for better fiber/matrix adhesion.	[82]
Jute reinforced polyester resin	Hand lay-up	Composite samples were manufactured through hand lay-up techniques and were tested for different mechanical properties.	Values of tensile strength, elongation at break and Young's modulus for untreated jute composite are 12.61 MPa, 20.96%, 84.63 MPa, respectively.	7% NaOH treated sample exhibited highest values of tensile strength, elongation at break and Young's modulus with increase of 48.69%, 87.5%, and 62.94% respectively from untreated sample. NaOH treatment makes surface rough, improving fiber/matrix adhesion which enhances mechanical properties.	[83]

4. Surface Treatments of Jute Fiber

Many surface-modification techniques are used to improve physical, chemical, and mechanical properties. Surface modification techniques help in improving adhesion between fiber and matrix, reducing water absorption, and enhancing fire-resistance properties [84,85]. These modification techniques are roughly divided into two groups—chemical and physical surface modification techniques. It is noted that not all techniques are eco- and environmentally-friendly. Hence, before using any techniques, environmental hazards must be kept in mind. For some applications, these modification techniques may not be suitable, such as in the food packaging industry. Chemical surface-modification techniques involve acetylation treatment, silane, alkaline treatment, enzymic treatment, succinic, and maleic anhydride grafting. Whereas physical modification techniques involve ultraviolet radiations, gamma irradiation, electron beam irradiation, plasma and corona treatment [86,87]. Most chemical modification techniques are discussed in fiber modification techniques, while physical modification techniques will be discussed below. Table 3 shows various surface modification methods.

Physical Modification Methods

Physical modification techniques are widely employed to improve the different properties of jute fiber in order to compete with synthetic counterparts. Gamma radiation is the process of deposition of energy in cellulose, which is achieved through a process called Compton scattering. After energy deposition, macro-cellulosic radicals are produced through the localization of deposited energy. These cellulosic radicals are responsible for improving different mechanical and physical properties of jute fiber [88]. Many studies are available on the gamma radiation of jute fiber, which focuses on the improvement of jute through various gamma doses. Khan et al. [89] studied the effects of gamma radiation for jute fiber reinforced polypropylene composite on various mechanical properties, with gamma radiations doses between 250–1000 krad. Results depicted that pretreatment of jute fibers and polypropylene with 500 krad of gamma radiation improved mechanical properties. Properties like tensile strength, impact strength, and bending strength improved by 27%, 73%, and 27%, respectively. Water intake for treated jute fiber composite was 6.97% as compared to 9.85% for untreated jute fiber composite. Gamma radiations can start scission and breaking of chemical bonds resulting in smaller polymeric molecules. The structure of matrix polymer is changed with the joining of these small molecules through cross-linking. Gamma radiations can increase active sites in matrix material. All these processes induced by radiations contribute in increasing mechanical properties of composite. Islam et al. [90] used gamma radiation for the surface modification of 50 wt % of treated jute, with 50–1000 krad of gamma radiation. An increase of 45% in tensile modulus and a 5% increase in tensile strength were observed in this experiment. Gamma treatment improved bond strength by providing more active site for better fber/matrix adhesion and induced cross-linking in matrix material. The jute fiber composite showed the best mechanical properties at a dose of 500 krad.

The UV radiation technique is another method for improving the mechanical properties of jute fibers. UV radiation techniques also help in cross-linking between the polymeric matrix and fiber. Khan et al. [88] studied the effects of gamma radiations alongside the effects of pre-irradiated fibers with UV radiations prior to gamma radiations. Gamma radiations improved different mechanical properties. Of all jute composites used in the experiment, composites with 38% jute content showed better mechanical properties. Tensile strength was increased by 108%, bending strength by 58%, bending modulus by 211%, and tensile modulus by 138%, as compared to a pure polymeric film. Tensile and bending strength increased up to 500 krad due to the formation of free radicals, increasing the degree of cross-linking. Tensile and bending strength decreased after 500 krad of radiation, mainly due to degradation of cellulose backbone [91]. Similarly, samples pre-irradiated with ultraviolet radiation before gamma treatment exhibited even better properties. A 15% increase in bending strength and a 19% increase in tensile strength were observed from the samples without UV treatment. There was a 150% increase in tensile strength and a 90% increase in bending strength in comparison with pure polymeric film. Abdullah-Al-Kafi et al. [55] studied the effects of UV radiation on jute/glass fiber composites where the 25 wt % treated jute exhibited better mechanical properties of all the samples. The UV radiated jute/glass fiber composite had increased tensile modulus by 33%, tensile strength by 70%, and tensile modulus by 33% compared with untreated composite. The increase in mechanical properties was attributed to an increase in cross-linking due to a phenomenon called photo-cross-linking, generating a high number of active sites. After a certain dose of UV radiation, mechanical properties show decrement in these mechanical properties mainly due to an opposite phenomenon—photodegradation. In photodegradation, the main cellulose chain starts to degrade and even polymer may degrade at high doses [91]. Plasma treatment was also found to be effective in improving the mechanical properties of jute fibers. In plasma treatment, new polar groups, or even polymer layers are introduced, helping the fiber to form a better adhesion with the matrix through covalent bonds. Seki et al. [92] studied the effects of oxygen plasma treatment on jute fibers. In this treatment, jute fibers were treated with radio frequencies and low frequencies through the use of different reactors. The treatment improved different mechanical properties, such as inter-laminar shear strength which was improved to 19.8 MPa for low frequency and 26.3 MPa for radio frequency oxygen plasma treatment compared to untreated jute fiber where the inter-laminar shear strength was around 11.5 MPa. This treatment also improved both flexural and tensile strengths. Radiofrequency oxygen plasma treatment was found to be much better than low frequency plasma treatment. Oxygen plasma improves fiber/matrix adhesion and interlocking due to an increase in roughness and by the fact that plasma treatment removes cellulose and hemicellulose, leaving lignin behind on the fiber surface which contributes in increasing fiber/matrix adhesion [93]. Electron-beam irradiation is an eco-friendly, clean and energy saving process to improve the surface properties of fibers, composite, films, and polymers [94]. Ji et al. [95] studied the effects of electron-beam irradiation on the mechanical and physical properties of jute fibers; 0–100 kGy (kiloGray) of electron-beam doses were used. In this study, jute fibers showed better thermal stability at the optimum dose of 10 kGy due to increased fiber/matrix adhesion.

5. Fiber Modification Techniques

5.1. Improvement in Jute Fiber and Matrix Adhesion

The mechanical properties of any natural composite usually depend upon the dispersion of fiber in a polymeric matrix. To ensure better mechanical and physical properties, there must be a proper adhesion between fiber and matrix. A strong adhesion between fiber and matrix will help with strong interfacial bonding [96]. A lack of proper adhesion between fiber and matrix can lead to a decrease in mechanical characteristics, such as a decrease in strength [97]. The hydrophilic nature of fiber and the hydrophobic nature of matrix is also one of the main reasons for poor adhesion. This difference in nature leads to a poor stress transfer between matrix and fiber, which will further lead to different problems like the cracking of the composite and a reduction in the different properties of composites [98]. It has been presented in many studies that different techniques can be used to overcome the issues related to adhesion. The surface-modification techniques of fibers seem to achieve good results in the improvement of fiber–matrix adhesion. Many chemical and physical surface-modification techniques have been established so far as part of attempts to overcome this issue [87]. Pukanszky [99] explained a modal for the quantitative description for the reliance of tensile properties of composites on parameters such as polymer/filler adhesion and geometry. It was concluded from analysis that the ultimate tensile properties were influenced by polymer/filler adhesion and interfacial interactions. Interfacial specific surface area, surface modification, aggregation, filler and matrix properties influence the strength of composite. The proposed modal also unfolded faults and imperfections in composite such as voids, aggregation, and dewetting. All the faults and imperfections, which are directly related to the failure of composite i.e., initiation and propagation, will have impact on the tensile properties of composite. Smaller filler particles have better matrix/filler adhesion but aggregation leads to the development of the failure site, while large filler particles have poor matrix/filler which will lead to dewetting and cavity formation. Cavities will act as failure initiation sites. Above certain critical value of filler, matrix discontinuity will increase, leading to the brittle fracture of the composite.

Liu et al. [100] treated jute fibers with an alkali treatment (using NaOH) and Maleic anhydride-grafted polypropylene (MPP) emulsion. This surface treatment method was effective in improving the performance of jute fibers in composites by increasing fiber/matrix adhesion. The method served to modify the jute fiber surface along improving fiber dispersion and increased the mechanical strength of the jute fibers, while the alkali treatment enhanced the removal of waxes and fatty constituents. The flexural, tensile, and impact strength of jute fibers were improved due to these treatments. Both of these methods were promising, showing an improvement of interfacial bonding due to an increase in adhesion between jute fiber and matrix. MPP is commonly utilized as a coupling agent to improve fiber–matrix adhesion. Combining alkali treatment and MPP show even better results in improving fiber–matrix adhesion [101]. Mohanty et al. [86] used dewaxing, alkali treatment, cyanoethylation, and grafting for the surface modification of jute fibers. Mechanical properties, such as tensile and impact strength, were also improved due to these surface modifications, along with the improvement in fiber/matrix adhesion. Mwaikambo et al. [102] treated jute fibers with an alkali solution of NaOH. This surface technique altered the surface topology and crystallization of fibers along with enhancing the fiber/matrix adhesion. Alkali treatment made the surface rough, increasing the interlocking strength between fiber and matrix. In addition, mechanical properties were also improved due to alkalization and crystallization. Corrales et al. [103] chemically modified jute fibers by treating them with a derivative of fatty acid—oleoyl chloride. This chemical method helped in modifying the hydrophilic nature of the fiber by which fibers were left with an olefinic deposit, which resulted in improved fiber/matrix adhesion. Basak et al. [104] studied the effects of temperature on silane treatment for jute fiber surface modification. It was concluded that a high-temperature silane treatment significantly improved the mechanical properties of jute fiber when compared to a low-temperature treatment. Additionally, silane treatment helped further in improving adhesion between jute fiber and matrix. Thakur et al. [105] also used silane treatment with the help of a silane coupling agent. The silane coupling agent was found to be effective in improving the physio-chemical properties of fibers, as well as being effective for surface modification. Silane treatment increased flexural, tensile strength and Young's modulus of fibers up to 30% along with significant improvements in fiber–matrix adhesion [106]. Battegazzore et al. [107] used an interesting layer-by-layer (LBL) assembly technique to modify hemp fibers in a composite. LBL technique has been used to increase fiber/matrix adhesion through nanostructured coatings of interphase materials. LBL promotes better fiber/matrix adhesion, mass transfer, and ultimately the mechanical properties of the composite. Chitosan and sepiolite nanorods were used in composites as interphase materials and deposited LBL via water-based electrolysis. LBL deposition improved the moisture resistance of fibers, made the surface smooth, and produced nanotexturing. Nanotexturing improved fiber/matrix adhesion in the composite. A significant increase in elastic modulus and tensile strength was observed. The results of some surface treatments are also reported in Table 3.

5.2. Moisture Absorption Properties

The hydrophilic nature of natural fibers is one of the major hindrances faced when seeking to improve the mechanical properties of natural composites. The reason for this moisture absorption property is the presence of hydroxyl groups [69]. Moisture absorption can result in a reduction of the mechanical and physical properties, changing the dimensions [37]. In fact, the water adsorption property remains the chief hindrance in possible uses and beneficial applications. Water is absorbed by capillary actions and will also incorporate into any micro cracks and void spaces present in a bio-composite [108]. Water absorption is carried out through two methods—diffusion and percolation methods [109,110]. Natural fibers need to be altered physically and chemically in order to overcome this problem of moisture absorption. Different studies are available to help solve the problem of moisture absorption by giving rise to hydrophobic properties in natural fiber [34]. The durability of natural fiber composites has been at stake due to this absorption property, while the physical and mechanical properties of jute fiber composites have received more focus, meaning that little work has so far been done to improve the efficiency of jute fibers in hygroscopic environments. Enzyme grafting is an important technique for increasing the hydrophobic behavior of natural fiber. Grafting usually involves creation of reactive radicals through lignin oxidation. These reactive radicals act as grafting sites for oxidized or non-oxidized molecules of the choice [111,112]. Liu et al. [112] studied the effects of grafting dodecyl gallate on a jute fiber surface through HRP mediated oxidative polymerization techniques. The main goal of the research was to increase the hydrophobic characteristics of jute fibers. FTIR, SEM, and TGA confirmed surface modification through grafting, while it was observed that a hydrophobic group was introduced by a grafting agent. Hydrophobic nature was also tested through wetting time and contact angle tests. In doing so, the hydrophobic nature of jute fiber was found to be increased after surface modification. Enzymatic bonding has been studied for grafting functional molecules to lignocellulosic materials mediated by laccase. Enzymes such as laccase are emerging mediator catalysts for enzymatic grafting, having an environment friendly nature and moderate working conditions [113,114]. Dong et al. [114] studied the hydrophobic nature of jute fibers via laccase-mediated dodecyl gallate enzymatic grafting, together with exploring the feasibility of this method. Grafting and surface modification was confirmed by FTIR, SEM, XPS, and AFM, while an increase in the hydrophobic nature of jute fiber was confirmed through wetting time and the contact angle test. This research confirmed an increase in the hydrophobic nature of jute fibers after enzymatic graft surface modification, which also proved to be eco-friendly and cost-effective. In addition, surface modification increases certain mechanical properties of jute composite. Hu et al. [115] studied the different stages of moisture absorption in jute fiber/PLA composites. It was found that moisture is absorbed in three stages, including a short and abrupt moisture absorption stage, a slow and constant moisture absorption stage, and the fastest and most abrupt moisture absorption stage. A long exposure of fibers to a moisture environment would decrease their mechanical properties along with the degradation of the composite, although different types of coating can help to reduce water absorption in fiber. Many defects were found, including micro-cracks, and pore and surface relaxation during aging. Hong et al. [116] found that maleic anhydride decreased the hygroscopic behavior of jute fibers through surface modifications and by increasing the compatibility between the jute fiber and the matrix. Maleic anhydride increased the covalent bonds along with van dar Waals forces, at interfaces between the jute fiber and the matrix. Some mechanical properties were also improved due to the surface modification.

5.3. Thermal Degradation and Fire Resistance Properties

Flammability and thermal degradation are some of the properties of natural jute fibers that have limited the use of fibers in a vast range of applications. Flammability and combustion of any natural fiber based composite depend on factors such as composite nature, polymeric matrix, natural fibers, moisture content, thermal properties, density, and structure. Due to flammability, thermal degradation, and fire issues, natural jute fibers have not found their way into high-temperature applications [85]. It is important to understand the flammability of both fiber and polymeric matrix material. Polymeric matrix undergoes thermal and thermal oxidative decomposition during combustion. Combustion results in production of heat, dense smoke, and volatiles. These volatiles include carbon monoxide, hydrocarbons along with non-combustible and non-flammable gases such as carbon dioxide, hydrogen halides etc. Each polymer produces different volatiles and these volatiles are dependent on chemical nature of polymer. Volatiles produce free radicals which are involved in the decomposition and burning of a polymer. Final products, rate of decomposition, and the decomposition mechanism are not only dependent on the chemical nature but also on the physical properties of polymers. Physical properties such as glass-transition, decomposition, and melting temperatures impact the decomposition mechanism. At these temperatures, the polymer goes through a phase-transition which influences physical properties such as viscosity, modulus, density, and thermal conductivity. Char formation is another important aspect to gauge the fire and thermal decomposition of polymeric matrix. Highly cross-linked polymers or polymers which undergo cross-linking during decomposition usually form char during combustion. Char prevents heat from underlying polymer layers, acting like a heat barrier. Polymers with char formation characteristics have low flammability and high fire resistance. These polymers are, therefore, more desirable for polymeric matrix materials [117–119]. Natural fibers have decomposition temperatures less than glass transition temperatures and are called non-thermo plastics. Natural fibers have poor heat and flame resistance. Thermal degradation of natural fiber involves desorption of moisture, cellulose chains cross-linking, and the formation of volatiles, char, tar, and gases. Fibers with high cellulosic content have poor flammability and thermal degradation properties. Natural fiber is composed of cellulose, hemicellulose, lignin, pectin, waxes, and oil based materials. These all constituents take part in poor fire resistance and low temperature degradation. Temperature plays a vital role in the thermal stability of natural fibers as it controls thermal expansion and contraction and moisture sorption [120]. Low thermal stability increases the chances of cellulose degradation with the release of volatile compounds directly effecting mechanical and physical properties of natural fiber based composites [26]. Cellulose usually decomposes between the temperature range of 260–350 °C, hemicellulose decomposes in between 200–260 °C, lignin starts decomposing at 160 °C and continues up to 400 °C [121]. Natural composite usually degrades completely around 400–500 °C with the release of heat and toxic compounds [122]. Burning can produce combustible gases, toxic compounds, non-combustible gases, char and smoke [123]. Differences in chemical constituents of natural fiber can change thermal degradation, fire resistance, and flame properties of natural fiber [124]. Fiber structure and fiber orientation play an important role for the determination of flammability properties [125]. The presence of ash and silica can increase fire resistance. Cellulose based materials are easily burnt in the presence of oxygen.

All natural fibers are required to pass certain tests regarding fire resistance and thermal degradation in order to be used in practical applications. Jute fibers will generate thick black smoke upon catching fire, which can have deadly consequences. To ensure smooth operation and safety, a natural jute fiber composite is needed to pass the safety test [126]. It is therefore important to understand the underlying chemistry of jute fiber in order to overcome this issue. The poor fire-resistance properties of jute fibers has kept them away from numerous applications, such as those vital for aerospace and transportation sectors. A very small number of studies have been carried out so far to improve the fire resistance of jute and natural fibers [119]. On the other hand, due to extensive research on synthetic composites, many methods have been formulated to overcome poor fire resistance. Temperature plays an important role in the thermal stability of jute fibers. A higher temperature can lead to the degradation of jute fiber and composite with decreased mechanical and physical properties [6]. Sinha et al. [127] modified the surface of jute fiber using 5% of NaOH, with jute fibers being treated at room temperature. From DSC analysis, it was clear that the thermal stability of jute fibers decreased after treatment. The reason for the decrease in thermal stability could be down to the close packing of cellulose with resin. A long exposure time in an alkali treatment would lead to the removal of hemicellulose which will, in turn, decrease the thermal stability of jute fiber.

Nam et al. [128] modified fibers and performed silane and alkali surface treatments to increase fiber–matrix adhesion, looking to achieve a better thermal stability of jute fiber composites. Surface treatments were found to decrease weight loss during TGA analysis while increasing the thermal stability of fiber. Combustion phenomena take place when natural fibers come into contact with fire or even heat. The combustion of natural fibers depends on many factors, like the amount of oxygen, the flow of gases within the combustion area, and the heat generated during combustion [85].

Various techniques are employed to increase fire resistance in jute fibers. These techniques include the introduction of fire-resistant methods during processing [129], including coating with fire-resistance materials, the introduction of non-flammable resins, different types of polymers, a range of binders, the insulation of composite, and the introduction of nanoparticles [130]. The particle size of fire-resistant material is also important for fire-resistance properties. Some studies have even tried to use organic fire-retardant materials but, due to toxicity issues, they are not preferred. Some inorganic and polymeric materials are helpful for increasing the fire resistance of natural and jute fibers [131]. Fatima and Mohanty [132] studied effects of natural rubber and fire-retardant on jute composites. Jute composites with 5% natural rubber and 1% sodium phosphate as a fire-retardant exhibited least smoke density and ability to self-extinguish. Apart from jute based composites, Battegazzore et al. [133] reported interesting results for a phosphorus-based fire-retardant—ammonium dihydrogen phosphate (ADP). ADP has been reported numerous times in different research studies as a fire-retardant. Fire properties were studied for rice husk particles and hemp fiber boards. ADP protected material from heat flux by forming a carbonaceous layer preventing underlying material from oxygen, heat, and mass transfer. High content ADP samples showed the best fire-resistant properties. Matko et al. [134] studied the flame retardancy for different starch, polyurethane, and polypropylene based bio-composites. Starch based composites, with the addition of diammonium phosphate as a fire-retardant, had increased fire resistance and were more efficient than other polymers. It was concluded that increased fire resistance was attributed due to the presence of the polyol characteristic of matrix polymer and the introduction of a flame retardant. Similarly, the addition of aluminum trihydrate as a fire-retardant can delay ignition time and reduce peak heat release values [135].

Figure 4 illustrates the different phases of fire in natural fiber.

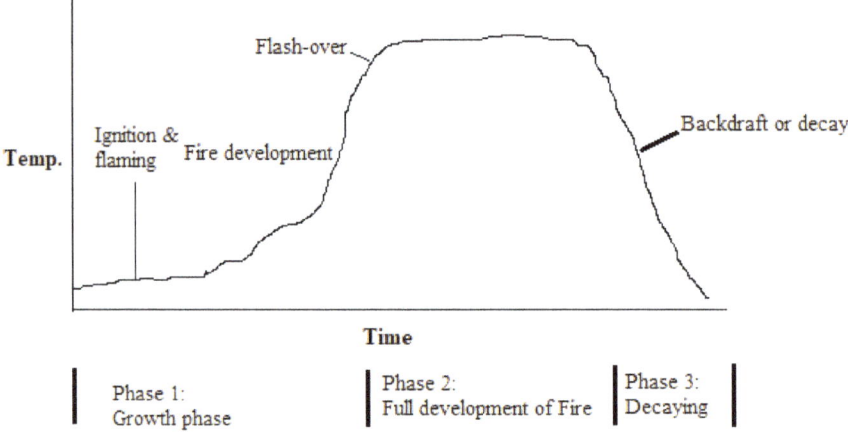

Figure 4. Different phases of fire in natural fiber. Redrawn with permission from Schartel, Hull [136] and Väisänen [3].

6. Processing Methods for Jute-Based Composites

Currently, bio-composites are being manufactured through conventional techniques used for production of synthetic composites. These techniques involve molding, resin transfer, compression molding, extrusion, injection, hand lay-on, spray lay-on, filament winding, and the pultrusion method [137]. These techniques have been formulated through years of industrial and research experience. Researchers have added different modifications to these techniques and even new techniques have developed but there is still capacity for much improvement to produce cost-effective and defect-free bio-composites. Bio-composites with little changes in process can be fabricated through these processes [138]. The processing route must be carefully selected to ensure proper dispersion, orientation, and aspect ratio of the fibers in the composite for desired applications [139]. The manufacturing route selection procedure also includes the consideration of the final design of the composite, size and shape, properties of raw materials, process speed and the overall cost [29]. A high aspect ratio along with uniform dispersion give rise to exceptional mechanical properties. Factors effecting manufacturing processes are moisture, fiber type, fiber contents, and fiber orientation.

These factors also influence the final properties of composites [29]. Moisture can significantly affect processing parameters and final properties. The adequate drying of fibers is necessary before proceeding for processing. Evaporation of water moisture during processes due to high temperature can cause bubbles in final products and increase porosity of bio-composites which can deteriorate mechanical properties [140]. Different fiber modification techniques can also be used to reduce moisture content [138]. Fiber type and its contents are very important for efficient and proper processing. In addition, fiber length, aspect ratio. and compounds like silicates present in fibers have great influence on processing [141]. Temperature is the most important factor that needs to be considered in the processing of bio-composites. Natural fibers usually have a small operating window for processing. Higher temperatures can lead to the degradation of fibers [12]. For the commercial production of bio-composites apart from an abundance of natural fibers, high efficiency and easy processability of these processing techniques are required [142]. Natural bio-composites need to have good structural and functional constancy during storage, in service, and during their final environmental degradation in order to compete with synthetic composites [2]. The main advantages of natural fiber processing are that these natural fibers cause minimal damage to tools used in process. Natural fibers are less abrasive than synthetic fibers and ultimately will cause less wear to tools and machines used in processing [143]. There are a few factors which must be kept in mind during processing. Stresses induced during processing can cause premature solidification of the melt. The final shape can undergo up to 8% shrinkage upon solidification [144]. Agglomeration is another problem encountered during processing. This problem is caused due to excess fiber use. Similar fibers cling to each other and this agglomeration can cause serious effects on the final mechanical properties of the composite [12]. High viscosity of melt effects process speediness and the uniformity of final products. The uniformity of the final product is also affected by the length of fibers [145]. Problems associated with processing techniques can be overcome through the addition of many additives but these additives will increase process cost. While additives may solve some problems, they can also create new problems in processing. A number of methods can be found for manufacturing jute fiber-based bio-composites, with key final properties helping to select the best possible option. One of the driving factors during processing is a homogenous distribution of fibers in the matrix. A homogenous distribution of fibers ensures better mechanical properties, while conditions like processing temperature are mainly dependent on a polymeric matrix [146]. Additionally, many methods have been developed for the manufacturing of bio-composites.

6.1. Hand Lay-Up Technique

The Hand Lay-Up Technique is a widely used method for the manufacturing of different bio-composites. In this method, the fibers are placed in the mold and resin is applied on fibers with the help of rollers. For curing, a vacuum technique is usually used afterward. Some of the main advantages of this process are its simplicity, low-processing cost, and the ability to manufacture complex designs; while the process is also known for having a long processing time and being labor intensive, which are the few disadvantages associated with the process [147,148]. During the fabrication of jute fiber composites, this process requires up to 400% more resin than fibers which is essentially not feasible for economic purposes. Along with excessive resin consumption, the process may require a pre-treatment of fibers, adding more cost to the process. Abdullah-Al-Kafi et al. [55] manufactured a jute and glass fiber composite by employing a hand lay-up technique, from which the composite with 25% of jute fibers showed better mechanical properties. Jute and glass fiber proportions were 1:3. Chaudhary et al. [21] fabricated a jute/hemp/flax epoxy reinforced hybrid composite using the hand lay-up technique. 8% jute, 9% hemp, and 8% flax fibers by weight were used for the fabrication of the hybrid composite.

6.2. Resin Transfer Molding

The process is considered to be better than hand lay-up techniques in terms of the quality of composites produced. Resin transfer molding is cost effective method with high production rate. First of all, the injection of resin material is carried out in the closed cavity mold after placement of fibers. Low vacuum pressure is applied for curing, which will be completed in 30–60 min due to vacuum. The fibers are placed in the mold prior to injection of resin. Then, after injection, fibers will impregnate in resin. Process parameters, such as injection pressure, vacuum pressure, fibers percentage, and temperature, affect the final mechanical and physical properties of the composite. This method is known for a high manufacturing rate and is ideal for the manufacturing of complex shapes. Mostly low-viscosity resins are used in this process [6,149]. Edge flow is observed in composite due to distance between fiber perform and mould cavity. Edge flow can easily disrupt the smoothness and uniformity of the flow pattern, leading to poor wetting of fibers. Edge flow can lead to defects during mould cavity filling which will lead to dry spot and spillage. The velocity of resin can vary from point to point due to non-uniformity and rough structure of fibers. This will cause voids to form which have adverse effects on the mechanical properties of the composite [150].

6.3. Pultrusion

Pultrusion is the best method for manufacturing composites with a thin cross-section; being a continuous manufacturing method, it is used to manufacture mats, ropes, yarn, etc. During the manufacturing of jute-based polymers and composites, jute fibers are impregnated in a resin material and then the material is passed through a hot die. In this process, it is very hard to keep fiber orientation constant. The process is known for the manufacturing of thin cross-sectional shapes and complex geometries with an allowance of high automation. Jute-based products that are manufactured via this process contain up to 70% of jute fibers. As a comparison, jute composites fabricated through this process have better mechanical properties, better electric insulation, and better corrosion resistance [6,151]. Akil et al. [152] studied a jute/glass fiber polyester hybrid composite. The hybrid composite was fabricated through pultrusion. The pulling speed and the die temperature were 180 mm/min and 85 °C, respectively, for pultrusion. The jute to glass fiber ratio was 50:50 by volume and fibers to matrix ratio was 70:30.

6.4. Extrusion

Extrusion is the most popular technique used by the plastics industry, being a technique that is known to offer a uniform mixing of all components. This method is useful for manufacturing composites where the orientation of fibers is not important, as the fibers are randomly distributed. In this process, single or twin screws are used which can rotate in clockwise and anti-clockwise directions. A single-screw extruder has low mixing effects and is used in applications where less mixing of fiber and matrix material is required. A twin-screw extruder is known for a great mixing effect and high-thrust forces, working to distribute fibers uniformly throughout the composite. With extrusion, ballets are usually fed into a heated chamber with feed screws to process molten mixture and, during the extrusion process of jute composite, up to 40% of jute fibers are mixed in a polymeric matrix. After this process, the final product can be subjected to post-processing techniques for higher quality. Figure 5 shows different fabrication methods [5,6,18].

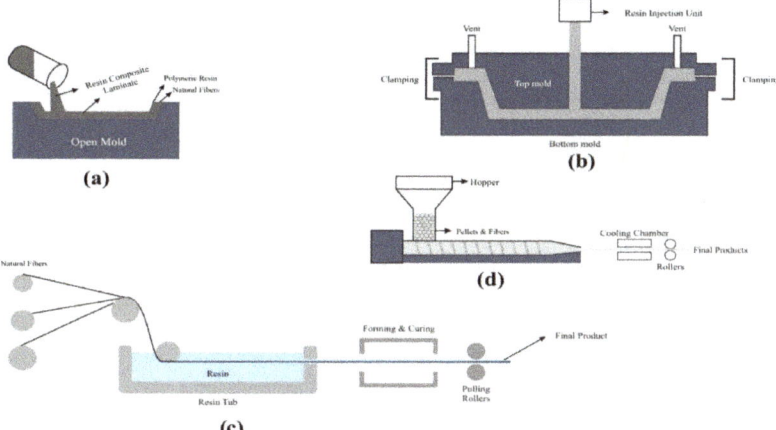

Figure 5. Different fabrication methods (**a**) Hand layup (**b**) Resin transfer molding (**c**) Pultrusion (**d**) Extrusion.

7. Hybrid Jute Bio-Composite

Natural fiber composites encounter various problems related to mechanical and physical properties. To overcome the shortcomings of natural fibers, synthetic fibers are introduced in composite matrix to make it a hybrid composite. Hybrid composites are simply the combination of two or more fibers in which one fiber compensates for the shortcomings of the other fiber. Usually, both natural and synthetic fibers are used to make hybrid composites. Hybrid bio-composites can also be termed as partial bio-degradable composites. Hybrid bio-composites may include non-degradable polymers with a combination of synthetic and natural fibers as fillers. Normally hybrid composites are manufactured through conventional fabrication techniques such as compression molding [2]. Hybridization of composites is generally classified into two types—intralaminate and interlaminate. In intralaminate hybridization, natural and synthetic fibers are intertwined together in one single layer while interlaminate is the deposition of different distinguishable fiber layers [153]. A fine balance between cost and performance is achievable through felicitous designing of hybrid composites. The physical and mechanical properties of hybrid composites are dependent on fiber content, length and orientation of each fiber, fiber/matrix bonding, and the failure strain of fibers. While designing and fabricating hybrid composites, the selection of suitable fibers and their properties are of utmost importance. The usefulness of a hybrid composite is determined by the physical, chemical, mechanical properties, and compatibility of fiber/matrix materials [4]. The addition of glass fibers helps in improving the shortcomings associated with natural fibers. Properties like elongation at break, impact strength, tensile strength, and Young's modulus in bio-composites can be enhanced with the hybridization of synthetic fibers. Hybridization reduces the water absorption property of bio-composites. Hybrid composites have certain advantages including high impact resistance, high toughness, and high specific strength. Composite hybridization is classified into two groups—hybrid composites with natural fibers and hybrid composites with synthetic fibers.

7.1. Hybridization with Natural Fibers

Composites with a combination of two or more natural fibers in a polymer matrix is known as a hybridization with natural fibers. Some of the advantages associated with natural fibers are low cost manufacturing, renewable nature, eco-friendliness, light weight, and optimum mechanical properties. Due to all these advantages, hybrid natural fiber composites are currently being used in number of applications such as in the automobile and aerospace industries. In general, hybrid composites have low strength to weight ratios and are easy to fabricate and manufacture [154].

Boopalan et al. [155] studied the thermal, mechanical, and water absorption properties of jute and banana fiber epoxy hybrid composites. Flexural, tensile and impact strength were at maximum for the 50/50 weight ratio of jute and banana fibers in an epoxy hybrid composite. Similarly, for the same ratio of both fibers, the hybrid composite had more enhanced thermal properties and a reduction in water absorption. Akil et al. [152] studied water uptake in a jute/glass fiber polyester hybrid composite. The addition of glass fibers increased the resistance of the composite towards water absorption. Water absorption was found to be dependent on fiber content and stacking sequence. When the hybrid composites were subjected to moisture, a significant reduction in flexural and tensile properties was observed due to water absorption. Fiore et al. [156] studied the effect of stacking sequence and sodium bi-carbonate treatment on the mechanical properties of a hybrid jute/flax epoxy composite. Sodium bi-carbonate treatment significantly improved the quasi-static properties of the hybrid composite. Sodium bi-carbonate treatment helped in improving the mechanical properties in the flax based composite, while the jute-based composite showed a slight decrease in mechanical properties. Optimum mechanical properties can be achieved through hybridization.

Jawaid et al. [157] studied chemical resistance, void content, and tensile properties for a tri-layer epoxy reinforced jute/oil palm fiber hybrid composite. Chemical resistance tests were performed using different chemicals. The hybrid composite had slightly elevated chemical resistance than the pure composite. When oil palm fiber was used as skin layer in tri-layer composite, it had a greater void content of 8.6% due to fiber/matrix compatibility issues. The jute/oil palm fiber/jute tri-layer composite had greater tensile strength as compared to when the jute fiber layer was sandwiched between oil palm fiber layers. This is due to the fact that jute fibers are more compatible with epoxy resin. In another research study, Jawaid et al. [158] studied the mechanical properties of a jute/oil palm fiber hybrid composite. The hybrid composite had increased flexural strength and modulus compared to the pure palm fiber composite. Impact strength was lower for the hybrid composite than the pure palm fiber composite. Shanmugam and Thiruchitrambalam [159] showed that alkali treated hybrid jute and palmyra palm leaf fibers had increased properties comparable with synthetic fiber hybridization such as glass fibers. Akali treated jute/palm leaf fibers based hybrid composites had increased tensile and flexural properties. Fiore et al. [160] studied the aging resistance of jute–basalt bio-epoxy hybrid composites. Sandwiched hybrid laminates exhibited high aging resistance due to basalt layers protecting jute fibers from degradation.

7.2. Hybridization with Synthetic Fibers

Synthetic fiber hybrid composites have better mechanical properties than natural fiber composites. Synthetic fibers have better fiber/matrix adhesion which increase the overall mechanical properties of the composite. But high cost associated with synthetic fiber manufacturing has limited the use of these hybrid composites in various applications. Some of issues related to synthetic fibers hybrid composites are environmental issues, recyclability, biodegradability, and reusability [161,162]. Research is being carried out in the field of synthetic fiber hybrid composites to overcome all the shortcomings. Researchers are looking into improving properties of natural fiber composites with the inclusion of synthetic fibers. The hybridization of natural fibers with glass fibers significantly improves the mechanical properties of the composite [154,163].

Ahmed et al. [164] studied different mechanical properties for a jute/glass fiber reinforced polyester hybrid composite. Young's modulus increased with the increase in glass fiber content while the Poisson ratio decreased. This was due to more transverse strain and lower longitudinal strain in the jute fiber composite in comparison with the jute/glass fiber composite. Aquino et al. [165] studied the effects of moisture on different mechanical properties of a jute/glass fiber hybrid composite. Moisture content decreased mechanical properties such as tensile strength and Young's modulus. Moisture disrupts fiber/matrix adhesion with decrease in mechanical properties. Selver et al. [166] studied the effect of stacking arrangement on different mechanical properties of a jute/flax/glass fiber thermoset composite. The addition of natural fibers reduced overall density for both jute/glass and flax/glass fiber composites. Higher flexural strength was obtained when glass fibers were used as outer layers, sandwiching natural fiber layer.

8. Limitations of Jute Fiber

Despite a lot of research being carried out to make use of jute fibers in many practical applications, many limitations still need to be addressed to get the full benefit from them. Some shortcomings of jute fibers include the hydrophilic nature of the fiber, poor fiber–matrix adhesion, the poor dispersion of jute fiber in a matrix, low physical and mechanical properties, flammable properties, limitations in thermal properties, a short temperature window for processing, and a lack of processing techniques [84,85,167]. These are some of the limitations keeping jute fibers from many applications, especially in load-bearing examples. Different scientists and researchers have developed fiber-modification techniques to overcome these issues. Jute fiber is hydrophilic in nature and vulnerable to absorbing water from the external environment. Hydroxyl and the presence of polar groups in jute fiber are responsible for the absorption of excess moisture. The moisture absorption in jute fiber can cause it to swell, which can further lead to the cracking of bio-composites. This moisture is also believed to cause compatibility issues with fiber and matrix. Poor fiber–matrix adhesion, a decrease in interfacial bonding, and a decrease in mechanical properties are some of the consequences of moisture absorption [97,167]. The hydrophobic nature of matrix material and the hydrophilic nature of fibers cause poor fiber–matrix adhesion in bio-composites. Different chemical methods are currently being used to overcome issues related to the hydrophilic nature of the fibers. Jute fiber composite usually has low mechanical properties when compared with synthetic fibers such as carbon and glass fibers. Much focus is being given on improving these mechanical properties to increase the implementation of jute fiber composites and polymers [168]. Different modification treatments can be applied to achieve such a goal. Researchers have found these treatments to be useful in improving different mechanical and physical properties [169]. Jute fibers are a natural fiber, containing pectin, lignin, oils, and different waxes. These constituents are all highly combustible and flammable in favorable conditions, making jute fiber poorly resistant to fire. The flammability issue is one of the major hindrances affecting the implementation of these fibers in practical use. However, the addition of flame retardants seems to be improving the flammability properties of these fibers, but much research is still required if this problem is to be overcome effectively. A high concentration of cellulose in jute fiber makes it more susceptible to fire [124]. Thermal degradation is another problem associated with these jute fibers, which can easily be degraded around a temperature of 450 °C, which limits jute fiber for low-temperature applications. This degradation will lead to drastic changes in the mechanical and physical properties of fibers [41]. Amount of jute fiber constituents such as cellulose, pectin, oil, and waxes can vary from plant to plant and are due to different external environmental factors. These amounts effect both thermal and fire properties, while fiber direction and structure also play an effective role in changing these properties. The high-cellulose composition makes fibers less fire-resistant, while low lignin concentration gives better fire resistance [125]. A lack of processing techniques for jute-based bio-composites is another issue. Currently, with little modifications, conventional techniques used for the production of synthetic composites and polymers are favored. However, new processing techniques are being developed to ensure the smooth processing of jute-based bio-composites.

9. Applications

Jute fiber has a vast range of applications and become one of the most important fibers in the bio-composite industry. Jute fiber somehow has better mechanical and physical properties than other natural fibers. Countries like Sri Lanka, Bangladesh, Malaysia, and Indonesia offer a high supply of jute fiber plants, making it abundant in nature. Currently, the textile industry is the main user of jute fiber, which is used to make clothes, ropes, bedsheets, sacks, bags, shoelaces, etc. Significantly, jute fiber has also made its way to the automobile sector, where it is used to make cup holders, different parts of the dashboard, and door panels. In the USA, many big companies have been using natural fibers like jute, hemp, and flex for making different exterior and interior parts for vehicles [170]. Jute fibers can help different car manufacturing companies to reduce weight and to improve mileage. Many big companies like BMW and Mercedes are taking the initiative and investing in research and development to make the best use of natural fibers in their cars. Furthermore, jute fibers have found applications in packaging industries and are replacing synthetic fibers, as well as being used in cosmetics, in the medical sector, and even in paints industries for various diversified applications.

Many developed countries are themselves taking radical steps to incorporate natural fibers into different industries, aiming for a clean environment. Jute fibers are readily used in construction for the manufacturing of windows, doors, floor matting, partitions between rooms, and for ceilings. They are even being used to make chairs, tables, and different kitchen products. European countries are more concerned with environmental changes and governments are desperately trying to introduce natural fibers like jute fiber into practical applications. For this purpose, many government and private organizations have joined hands for the commercialization of these fibers. Governments are giving different incentives to industries for the use of natural fibers in the manufacturing of products [171]. In addition, jute fiber is employed in agriculture and consumer goods industries for various applications. The use of various natural fibers, including jute fiber, will continue to increase in the coming few years, offering lots of potential to be used in a number of applications. In addition, jute fibers are renewable and eco-friendly in nature [172]. Although jute fiber is considered as one of the most popular fibers, a lot of effort is still required to make the best use of it in different applications. Some of the main reasons for the slow commercialization of jute fiber include a lack of processing methods, low mechanical and physical properties in comparison with synthetic fibers, and the high costs associated with jute fiber composites [6,24]. Research studies are being carried out to make jute fiber a better fit for its potential applications. Due to a huge number of potential applications, many industries are taking a fresh interest in jute fibers. Asia has become a hub for the production of jute fiber and many new markets are emerging throughout the region. It is anticipated that the demand for jute fiber will increase significantly in the near future due to vast applications. Consumers are accepting jute fiber composites due to its excellent properties and preferring its products due to minimal environmental effects. Figure 6 shows the different applications of jute-based composites.

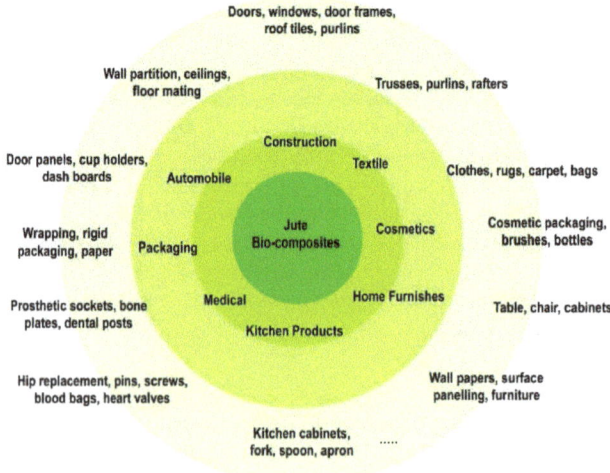

Figure 6. Applications of jute-based composites.

10. Conclusions

The enormous potential of jute-based composites to provide environmentally-friendly materials is the key driving force behind their fast development. The concept of bio-composites is not new, rather the term "green composites" has been tossed around by a number of studies before. Initially, high fabrication costs and a lack of synthesis methods have restricted the growth of bio-composites, but environmental concerns have raised their importance. Jute-based composites have attracted the attention of many researchers, together with research and development funds, due to their better physical and mechanical properties among all natural fibers.

This review depicts the different problems linked with an excessive use of synthetic composites. These problems include pollution, disposal problems, inertness, and the high-carbon emissions associated with their use. It is concluded that the matter of discontinuing and discouraging the manufacturing of synthetic composites is of absolute importance for protecting the environment from their hazards. In this work, we have presented a critical analysis and some key findings concerning recent jute-based bio and hybrid composites. These key findings include discussion of different fabrication techniques for jute-based composites, such as hand lay-up, resin transfer molding, pultrusion and extrusion, and detailed discussion regarding the physical, mechanical, and electrical issues, as well as flammability and moisture absorption properties. The structure of jute fibers has been discussed, describing all the constituents and the part they play in its properties. It was determined that micro-fibrils and lumen play vital roles in both the mechanical properties and low density. Different bonding structures were also discussed, explaining their roles in better mechanical properties. A few of the significant limitations of jute-based composites include poor fiber–matrix adhesion and the hydrophilic nature of the fiber. These limitations can be overcome through different physical–chemical and chemical-modification techniques. These modification techniques help in improving poor fiber–matrix adhesion, the hydrophilic nature, and the poor thermal properties of jute fibers. Different surface-modification techniques, such as alkali and radiation techniques, were also found to be useful in increasing the mechanical properties and minimizing the limitations of jute-based composites. Another way to overcome these limitations is the inclusion of synthetic fibers to make hybrid jute-based composites. Hybrid jute-based composites have better mechanical and physical properties. Hybrid jute-based composites have improved fiber/matrix adhesion, better resistance for moisture, and enhanced thermal properties.

This study also outlines the various applications of jute-based composites, such as textiles, construction, packaging, medical, cosmetic, and furniture industries. Due to the growing demand for jute-based composites, the applications are likely to include new fields and it is expected that their usage will be increased significantly, followed by their detailed research and development for industrial-scale processes.

Author Contributions: This review study was conceptualized with cooperation among all authors; M.A.A. and A.B. designed the review topic and performed literature search. M.Z. and M.T.M. handled the writing and editing of the manuscript. R.K. and A.B. assisted with the development of tables and figures from the data analysis.

Funding: This article was funded by the Deanship of Scientific Research (DSR) at King Abdulaziz University, Jeddah. The author, therefore, acknowledges with thanks DSR for technical and financial support.

Conflicts of Interest: The authors declare no conflict of interest.

References

1. dos Santos Rosa, D.; Lenz, D.M. Biocomposites: Influence of matrix nature and additives on the properties and biodegradation behaviour. *Biodegrad. Eng. Technol.* **2013**. [CrossRef]
2. Satyanarayana, K.G.; Arizaga, G.G.; Wypych, F.; Arizaga, G.G.C. Biodegradable composites based on lignocellulosic fibers—An overview. *Prog. Polym. Sci.* **2009**, *34*, 982–1021. [CrossRef]
3. Väisänen, T.; Das, O.; Tomppo, L. A review on new bio-based constituents for natural fiber-polymer composites. *J. Clean. Prod.* **2017**, *149*, 582–596. [CrossRef]

4. John, M.J.; Thomas, S. Biofibres and biocomposites. *Carbohydr. Polym.* **2008**, *71*, 343–364. [CrossRef]
5. Burrola-Núñez, H.; Herrera-Franco, P.J.; Rodríguez-Félix, D.E.; Soto-Valdez, H.; Madera-Santana, T.J. Surface modification and performance of jute fibers as reinforcement on polymer matrix: An overview. *J. Nat. Fibers* **2018**, 1–17. [CrossRef]
6. Iman, M.; Maji, T.K. Jute: An Interesting Lignocellulosic Fiber for New Generation Applications. *Lignocellul. Polym. Compos.* **2014**, *3*, 453–475.
7. Kozlowski, R.; Wladyka-Przybylak, M. Uses of natural fiber reinforced plastics. In *Natural Fibers, Plastics and Composites*; Springer: Berlin, Germany, 2004; pp. 249–274.
8. Satyanarayana, K.G.; Flores-Sahagun, T.H.; Bowman, P. Lignocellulosic Materials of Brazil—Their Characterization and Applications in Polymer Composites and Art Works. In *Lignocellulosic Composite Materials*; Springer: Berlin, Germany, 2018; pp. 1–96.
9. Welcome to the World of Jute and Kenaf—IJSG. Available online: http://www.jute.org/plant.htm (accessed on 27 June 2018).
10. Mir, R.R.; Rustgi, S.; Sharma, S.; Singh, R.; Goyal, A.; Kumar, J.; Gaur, A.; Tyagi, A.K.; Khan, H.; Sinha, M.K.; et al. A preliminary genetic analysis of fibre traits and the use of new genomic SSRs for genetic diversity in jute. *Euphytica* **2008**, *161*, 413–427. [CrossRef]
11. Li, X.; Tabil, L.G.; Panigrahi, S. Chemical Treatments of Natural Fiber for Use in Natural Fiber-Reinforced Composites: A Review. *J. Polym. Environ.* **2007**, *15*, 25–33. [CrossRef]
12. Pickering, K.; Efendy, M.A.; Le, T. A review of recent developments in natural fibre composites and their mechanical performance. *Compos. Part A Appl. Sci. Manuf.* **2016**, *83*, 98–112. [CrossRef]
13. Wang, P.; Chen, F.; Zhang, H.; Meng, W.; Sun, Y.; Liu, C. Large-scale preparation of jute-fiber-reinforced starch-based composites with high mechanical strength and optimized biodegradability. *Starch Stärke* **2017**, *69*, 1700052. [CrossRef]
14. Jahan, M.S.; Saeed, A.; He, Z.; Ni, Y. Jute as raw material for the preparation of microcrystalline cellulose. *Cellulose* **2011**, *18*, 451–459. [CrossRef]
15. Monteiro, S.N.; Lopes, F.P.D.; Ferreira, A.S.; Nascimento, D.C.O. Natural-fiber polymer-matrix composites: Cheaper, tougher, and environmentally friendly. *JOM* **2009**, *61*, 17–22. [CrossRef]
16. Suddell, B.C.; Evans, W.J.; Isaac, D.H.; Crosky, A. A survey into the application of natural fiber composites in the automotive industry. In Proceedings of the 4th International Symposium on Natural Polymers and Composites—ISNAPol, São Pedro, SP, Brazil, 10–11 April 2002; pp. 455–461.
17. Pickering, K.L. *Properties and Performance of Natural-Fibre Composites*; Elsevier: Amsterdam, The Netherlands, 2008.
18. Faruk, O.; Bledzki, A.K.; Fink, H.-P.; Sain, M. Biocomposites reinforced with natural fibers: 2000–2010. *Prog. Polym. Sci.* **2012**, *37*, 1552–1596. [CrossRef]
19. PressReader.com—Connecting People Through News. Available online: https://www.pressreader.com/philippines/manila-bulletin/20170619/281938837908620 (accessed on 11 May 2018).
20. Top Jute Producing Countries in the World. In: World Atlas. Available online: https://www.worldatlas.com/articles/top-jute-producing-countries-in-the-world.html (accessed on 11 May 2018).
21. Chaudhary, V.; Bajpai, P.K.; Maheshwari, S. Studies on Mechanical and Morphological Characterization of Developed Jute/Hemp/Flax Reinforced Hybrid Composites for Structural Applications. *J. Nat. Fibers* **2018**, *15*, 80–97. [CrossRef]
22. Abraham, E.; Deepa, B.; Pothan, L.; Jacob, M.; Thomas, S.; Cvelbar, U.; Anandjiwala, R. Extraction of nanocellulose fibrils from lignocellulosic fibres: A novel approach. *Carbohydr. Polym.* **2011**, *86*, 1468–1475. [CrossRef]
23. Vilay, V.; Mariatti, M.; Taib, R.M.; Todo, M. Effect of fiber surface treatment and fiber loading on the properties of bagasse fiber–reinforced unsaturated polyester composites. *Compos. Sci. Technol.* **2008**, *68*, 631–638. [CrossRef]
24. Roe, P.J.; Ansell, M.P. Jute-reinforced polyester composites. *J. Mater. Sci.* **1985**, *20*, 4015–4020. [CrossRef]
25. Biagiotti, J.; Puglia, D.; Kenny, J.M. A review on natural fibre-based composites-part I: Structure, processing and properties of vegetable fibres. *J. Nat. Fibers* **2004**, *1*, 37–68. [CrossRef]
26. Jawaid, M.; Khalil, H.A. Cellulosic/synthetic fibre reinforced polymer hybrid composites: A review. *Carbohydr. Polym.* **2011**, *86*, 1–18. [CrossRef]
27. Reddy, N.; Yang, Y. Biofibers from agricultural byproducts for industrial applications. *Trends Biotechnol.* **2005**, *23*, 22–27. [CrossRef]

28. Guillén, F.; Martínez, M.J.; Gutiérrez, A.; Del Rio, J.C. Biodegradation of lignocellu-losics: Microbial, chemical, and enzymatic aspects of the fungal attack of lignin. *Int. Microbiol.* **2005**, *8*, 195–204.
29. Faruk, O.; Bledzki, A.K.; Fink, H.-P.; Sain, M. Progress report on natural fiber reinforced composites. *Macromol. Mater. Eng.* **2014**, *299*, 9–26. [CrossRef]
30. Khalil, H.A.; Davoudpour, Y.; Islam, M.N.; Mustapha, A.; Sudesh, K.; Dungani, R.; Jawaid, M. Production and modification of nanofibrillated cellulose using various mechanical processes: A review. *Carbohydr. Polym.* **2014**, *99*, 649–665. [CrossRef] [PubMed]
31. Kalia, S.; Dufresne, A.; Cherian, B.M.; Kaith, B.S.; Avérous, L.; Njuguna, J.; Nassiopoulos, E. Cellulose-Based Bio- and Nanocomposites: A Review. *Int. J. Polym. Sci.* **2011**. [CrossRef]
32. Khalil, H.A.; Bhat, A.; Yusra, A.I. Green composites from sustainable cellulose nanofibrils: A review. *Carbohydr. Polym.* **2012**, *87*, 963–979. [CrossRef]
33. Mohanty, A.K.; Misra, M.; Hinrichsen, G. Biofibers, biodegradable polymers and biocomposites: An overview. *Macromol. Mater. Eng.* **2000**, *276*, 1–24. [CrossRef]
34. Akil, H.M.; Cheng, L.W.; Ishak, Z.M.; Abu Bakar, A.; Rahman, M.A. Water absorption study on pultruded jute fibre reinforced unsaturated polyester composites. *Compos. Sci. Technol.* **2009**, *69*, 1942–1948. [CrossRef]
35. Klemm, D.; Philipp, B.; Heinze, T.; Heinze, U.; Wagenknecht, W. General considerations on structure and reactivity of cellulose: Section 2.1–2.1. *Wiley Online Library* **2004**, *4*. [CrossRef]
36. Gopinath, A.; Kumar, M.S.; Elayaperumal, A. Experimental Investigations on Mechanical Properties of Jute Fiber Reinforced Composites with Polyester and Epoxy Resin Matrices. *Procedia Eng.* **2014**, *97*, 2052–2063. [CrossRef]
37. Célino, A.; Freour, S.; Jacquemin, F.; Casari, P. The hygroscopic behavior of plant fibers: A review. *Front. Chem.* **2014**, *1*. [CrossRef]
38. Fratzl, P. Cellulose and collagen: From fibres to tissues. *Curr. Opin. Colloid Interface Sci.* **2003**, *8*, 32–39. [CrossRef]
39. Roy, S.; Lutfar, L.B. 3—Bast fibres: Jute. In *Handbook of Natural Fibres*; Kozłowski, R.M., Ed.; Woodhead Publishing: Sawston, UK, 2012; pp. 24–46.
40. Jayamani, E.; Hamdan, S.; Rahman, M.R.; Bin Bakri, M.K. Comparative Study of Dielectric Properties of Hybrid Natural Fiber Composites. *Procedia Eng.* **2014**, *97*, 536–544. [CrossRef]
41. Yan, L.; Kasal, B.; Huang, L. A review of recent research on the use of cellulosic fibres, their fibre fabric reinforced cementitious, geo-polymer and polymer composites in civil engineering. *Compos. Part B Eng.* **2016**, *92*, 94–132. [CrossRef]
42. Hansen, C.M.; Björkman, A. The Ultrastructure of Wood from a Solubility Parameter Point of View. *Holzforsch Int. J. Biol. Chem. Phys. Technol. Wood* **2009**, *52*, 335–344. [CrossRef]
43. Rowell, R.M.; Han, J.S.; Rowell, J.S. Characterization and factors effecting fiber properties. *Nat. Polym. Agrofibers Based Compos.* **2000**, *2000*, 115–134.
44. Norgren, M.; Edlund, H. Lignin: Recent advances and emerging applications. *Curr. Opin. Colloid Interface Sci.* **2014**, *19*, 409–416. [CrossRef]
45. Kumar, M.N.S.; Mohanty, A.K.; Erickson, L.; Misra, M. Lignin and Its Applications with Polymers. *J. Biobased Mater. Bioenergy* **2009**, *3*, 1–24. [CrossRef]
46. Gurunathan, T.; Mohanty, S.; Nayak, S.K. A review of the recent developments in biocomposites based on natural fibres and their application perspectives. *Compos. Part A Appl. Sci. Manuf.* **2015**, *77*, 1–25. [CrossRef]
47. Mishra, V.; Biswas, S. Physical and Mechanical Properties of Bi-directional Jute Fiber Epoxy Composites. *Procedia Eng.* **2013**, *51*, 561–566. [CrossRef]
48. Thwe, M.M.; Liao, K. Durability of bamboo-glass fiber reinforced polymer matrix hybrid composites. *Compos. Sci. Technol.* **2003**, *63*, 375–387. [CrossRef]
49. Fu, S.-Y.; Xu, G.; Mai, Y.-W. On the elastic modulus of hybrid particle/short-fiber/polymer composites. *Compos. Part B Eng.* **2002**, *33*, 291–299. [CrossRef]
50. Fiore, V.; Scalici, T.; Sarasini, F.; Tirilló, J.; Calabrese, L. Salt-fog spray aging of jute-basalt reinforced hybrid structures: Flexural and low velocity impact response. *Compos. Part B Eng.* **2017**, *116*, 99–112. [CrossRef]
51. Ho, M.-P.; Wang, H.; Lee, J.-H.; Ho, C.-K.; Lau, K.-T.; Leng, J.; Hui, D. Critical factors on manufacturing processes of natural fibre composites. *Compos. Part B Eng.* **2012**, *43*, 3549–3562. [CrossRef]

52. Chollakup, R.; Tantatherdtam, R.; Ujjin, S.; Sriroth, K. Pineapple leaf fiber reinforced thermoplastic composites: Effects of fiber length and fiber content on their characteristics. *J. Appl. Polym. Sci.* **2011**, *119*, 1952–1960. [CrossRef]
53. Lodha, P.; Netravali, A.N. Characterization of interfacial and mechanical properties of "green" composites with soy protein isolate and ramie fiber. *J. Mater. Sci.* **2002**, *37*, 3657–3665. [CrossRef]
54. Mishra, H.K.; Dash, B.N.; Tripathy, S.S.; Padhi, B.N. A study on mechanical performance of jute-epoxy composites. *Polym. Technol. Eng.* **2000**, *39*, 187–198. [CrossRef]
55. Abdullah-Al-Kafi; Abedin, M.Z.; Beg, M.D.H.; Pickering, K.L.; Khan, M.A. Study on the mechanical properties of jute/glass fiber-reinforced unsaturated polyester hybrid composites: Effect of surface modification by ultraviolet radiation. *J. Reinf. Plast. Compos.* **2006**, *25*, 575–588. [CrossRef]
56. Ahmed, K.S.; Vijayarangan, S. Tensile, flexural and interlaminar shear properties of woven jute and jute-glass fabric reinforced polyester composites. *J. Mater. Process. Technol.* **2008**, *207*, 330–335. [CrossRef]
57. Ahmed, K.S.; Vijayarangan, S.; Rajput, C. Mechanical Behavior of Isothalic Polyester-based Untreated Woven Jute and Glass Fabric Hybrid Composites. *J. Reinf. Plast. Compos.* **2006**, *25*, 1549–1569. [CrossRef]
58. Gibeop, N.; Lee, D.; Prasad, C.; Toru, F.; Kim, B.S.; Song, J.I. Effect of plasma treatment on mechanical properties of jute fiber/poly (lactic acid) biodegradable composites. *Adv. Compos. Mater.* **2013**, *22*, 389–399. [CrossRef]
59. Zamri, M.H.; Akil, H.M.; Bakar, A.A.; Ishak, Z.A.M.; Cheng, L.W. Effect of water absorption on pultruded jute/glass fiber-reinforced unsaturated polyester hybrid composites. *J. Compos. Mater.* **2012**, *46*, 51–61. [CrossRef]
60. Goriparthi, B.K.; Suman, K.; Rao, N.M. Effect of fiber surface treatments on mechanical and abrasive wear performance of polylactide/jute composites. *Compos. Part A Appl. Sci. Manuf.* **2012**, *43*, 1800–1808. [CrossRef]
61. Rezaur Rahman, M.; Hasan, M.; Monimul Huque, M.; Nazrul Islam, M. Physico-mechanical properties of jute fiber reinforced polypropylene composites. *J. Reinf. Plast. Compos.* **2010**, *29*, 445–455. [CrossRef]
62. Boopalan, M.; Umapathy, M.J.; Jenyfer, P. A Comparative Study on the Mechanical Properties of Jute and Sisal Fiber Reinforced Polymer Composites. *Silicon* **2012**, *4*, 145–149. [CrossRef]
63. Joshy, M.K.; Mathew, L.; Joseph, R. Influence of Fiber Surface Modification on the Mechanical Performance of Isora-Polyester Composites. *Int. J. Polym. Mater.* **2008**, *58*, 2–20. [CrossRef]
64. Lilholt, H.; Lawther, J.M. 1.10—Natural Organic Fibers. In *Comprehensive Composite Materials*; Kelly, A., Zweben, C., Eds.; Pergamon: Oxford, UK, 2000; pp. 303–325.
65. Rowell, R.M.; Sanadi, A.R.; Caulfield, D.F.; Jacobson, R.E. Utilization of natural fibers in plastic composites: Problems and opportunities. *Lignocellul. Plast Compos.* **1997**, *13*, 23–51.
66. Stokke, D.D.; Wu, Q.; Han, G. *Introduction to Wood and Natural Fiber Composites*; John Wiley & Sons: Hoboken, NJ, USA, 2013.
67. Ramesh, M.; Palanikumar, K.; Reddy, K.H. Mechanical property evaluation of sisal–jute–glass fiber reinforced polyester composites. *Compos. Part B Eng.* **2013**, *48*, 1–9. [CrossRef]
68. Ramesh, M.; Palanikumar, K.; Reddy, K.H. Comparative Evaluation on Properties of Hybrid Glass Fiber-Sisal/Jute Reinforced Epoxy Composites. *Procedia Eng.* **2013**, *51*, 745–750. [CrossRef]
69. Kalia, S.; Kaith, B.; Kaur, I. Pretreatments of natural fibers and their application as reinforcing material in polymer composites—A review. *Polym. Eng. Sci.* **2009**, *49*, 1253–1272. [CrossRef]
70. Wambua, P.; Ivens, J.; Verpoest, I. Natural fibres: Can they replace glass in fibre reinforced plastics? *Compos. Sci. Technol.* **2003**, *63*, 1259–1264. [CrossRef]
71. Monteiro, S.N.; Lopes, F.P.D.; Barbosa, A.P.; Bevitori, A.B.; Da Silva, I.L.A.; Da Costa, L.L. Natural Lignocellulosic Fibers as Engineering Materials—An Overview. *Met. Mater. Trans. A* **2011**, *42*, 2963–2974. [CrossRef]
72. Satyanarayana, K.; Guimarães, J.; Wypych, F. Studies on lignocellulosic fibers of Brazil. Part I: Source, production, morphology, properties and applications. *Compos. Part A Appl. Sci. Manuf.* **2007**, *38*, 1694–1709. [CrossRef]
73. Dufresne, A. Cellulose-based composites and nanocomposites. In *Monomers, Polymers and Composites from Renewable Resources*; Elsevier: Amsterdam, The Netherlands, 2008; pp. 401–418.
74. Dittenber, D.B.; GangaRao, H.V. Critical review of recent publications on use of natural composites in infrastructure. *Compos. Part A Appl. Sci. Manuf.* **2012**, *43*, 1419–1429. [CrossRef]

75. Dicker, M.P.; Duckworth, P.F.; Baker, A.B.; François, G.; Hazzard, M.K.; Weaver, P.M. Green composites: A review of material attributes and complementary applications. *Compos. Part A Appl. Sci. Manuf.* **2014**, *56*, 280–289. [CrossRef]
76. Rathore, A.; Pradhan, M.; Pradhan, M. Hybrid Cellulose Bionanocomposites from banana and jute fibre: A Review of Preparation, Properties and Applications. *Mater. Today Proc.* **2017**, *4*, 3942–3951. [CrossRef]
77. Cheung, H.-Y.; Ho, M.-P.; Lau, K.-T.; Cardona, F.; Hui, D. Natural fibre-reinforced composites for bioengineering and environmental engineering applications. *Compos. Part B Eng.* **2009**, *40*, 655–663. [CrossRef]
78. Zini, E.; Scandola, M. Green composites: An overview. *Polym. Compos.* **2011**, *32*, 1905–1915. [CrossRef]
79. Tripathi, P.; Gupta, V.K.; Dixit, A.; Mishra, R.K.; Sharma, S. Development and characterization of low cost jute, bagasse and glass fiber reinforced advanced hybrid epoxy composites. *AIMS Mater. Sci.* **2018**, *5*, 320–337. [CrossRef]
80. Jha, K.; Samantaray, B.B.; Tamrakar, P. A Study on Erosion and Mechanical Behavior of Jute/E-Glass Hybrid Composite. *Mater. Today Proc.* **2018**, *5*, 5601–5607. [CrossRef]
81. Anand, P.; Rajesh, D.; Kumar, M.S.; Raj, I.S. Investigations on the performances of treated jute/Kenaf hybrid natural fiber reinforced epoxy composite. *J. Polym. Res.* **2018**, *25*, 94. [CrossRef]
82. Motaleb, K.A. Improvement of Mechanical Properties by Alkali Treatment on Pineapple and Jute Fabric Reinforced Polyester Resin Composites. *Int J. Compos. Mater.* **2018**, *8*, 32–37.
83. Athijayamani, A.; Thiruchitrambalam, M.; Natarajan, U.; Pazhanivel, B. Effect of moisture absorption on the mechanical properties of randomly oriented natural fibers/polyester hybrid composite. *Mater. Sci. Eng. A* **2009**, *517*, 344–353. [CrossRef]
84. Kozlowski, R.; Władyka-Przybylak, M.; Władyka-Przybylak, M. Flammability and fire resistance of composites reinforced by natural fibers. *Polym. Adv. Technol.* **2008**, *19*, 446–453. [CrossRef]
85. Mohanty, A.; Khan, M.A.; Hinrichsen, G. Surface modification of jute and its influence on performance of biodegradable jute-fabric/Biopol composites. *Compos. Sci. Technol.* **2000**, *60*, 1115–1124. [CrossRef]
86. Mohanty, A.K.; Misra, M.; Drzal, L.T. Surface modifications of natural fibers and performance of the resulting biocomposites: An overview. *Compos. Interfaces* **2001**, *8*, 313–343. [CrossRef]
87. Braga, R.; Magalhaes, P. Analysis of the mechanical and thermal properties of jute and glass fiber as reinforcement epoxy hybrid composites. *Mater. Sci. Eng. C* **2015**, *56*, 269–273. [CrossRef] [PubMed]
88. Khan, M.A.; Haque, N.; Al-Kafi, A.; Alam, M.N.; Abedin, M.Z. Jute Reinforced Polymer Composite by Gamma Radiation: Effect of Surface Treatment with UV Radiation. *Polym. Technol. Eng.* **2006**, *45*, 607–613. [CrossRef]
89. Khan, R.A.; Khan, M.A.; Khan, A.H.; Hossain, M.A. Effect of gamma radiation on the performance of jute fabrics-reinforced polypropylene composites. *Radiat. Phys. Chem.* **2009**, *78*, 986–993.
90. Islam, T.; Khan, R.A.; Khan, M.A.; Rahman, M.A.; Fernandez-Lahore, M.; Huque, Q.M.I.; Islam, R.; Lahore, H.M.F. Physico-Mechanical and Degradation Properties of Gamma-Irradiated Biocomposites of Jute Fabric-Reinforced Poly(caprolactone). *Polym. Technol. Eng.* **2009**, *48*, 1198–1205. [CrossRef]
91. Hassan, M.M.; Islam, M.R.; Shehrzade, S.; Khan, M.A. Influence of Mercerization Along with Ultraviolet (UV) and Gamma Radiation on Physical and Mechanical Properties of Jute Yarn by Grafting with 3-(Trimethoxysilyl) Propylmethacrylate (Silane) and Acrylamide Under UV Radiation. *Polym. Technol. Eng.* **2003**, *42*, 515–531. [CrossRef]
92. Seki, Y.; Sarikanat, M.; Sever, K.; Erden, S.; Gulec, H.A. Effect of the low and radio frequency oxygen plasma treatment of jute fiber on mechanical properties of jute fiber/polyester composite. *Fibers Polym.* **2010**, *11*, 1159–1164. [CrossRef]
93. Yuan, X.; Jayaraman, K.; Bhattacharyya, D. Effects of plasma treatment in enhancing the performance of woodfibre-polypropylene composites. *Compos. Part A Appl. Sci. Manuf.* **2004**, *35*, 1363–1374. [CrossRef]
94. Pang, Y.; Cho, D.; Han, S.O.; Park, W.H. Interfacial shear strength and thermal properties of electron beam-treated henequen fibers reinforced unsaturated polyester composites. *Macromol. Res.* **2005**, *13*, 453–459. [CrossRef]
95. Ji, S.G.; Hwang, J.H.; Cho, D.; Kim, H.-J. Influence of electron beam treatment of jute on the thermal properties of random and two-directional jute/poly(lactic acid) green composites. *J. Adhes. Sci. Technol.* **2013**, *27*, 1359–1373. [CrossRef]

96. Signori, F.; Pelagaggi, M.; Bronco, S.; Righetti, M.C. Amorphous/crystal and polymer/filler interphases in biocomposites from poly(butylene succinate). *Thermochim. Acta* **2012**, *543*, 74–81. [CrossRef]
97. John, M.J.; Anandjiwala, R.D. Recent developments in chemical modification and characterization of natural fiber-reinforced composites. *Polym. Compos.* **2008**, *29*, 187–207. [CrossRef]
98. Fiore, V.; Di Bella, G.; Valenza, A. The effect of alkaline treatment on mechanical properties of kenaf fibers and their epoxy composites. *Compos. Part B Eng.* **2015**, *68*, 14–21. [CrossRef]
99. Pukánszky, B. Influence of interface interaction on the ultimate tensile properties of polymer composites. *Composites* **1990**, *21*, 255–262. [CrossRef]
100. Liu, X. Surface modification and micromechanical properties of jute fiber mat reinforced polypropylene composites. *Express Polym. Lett.* **2007**, *1*, 299–307. [CrossRef]
101. Mohanty, A.K.; Drzal, L.T.; Misra, M. Novel hybrid coupling agent as an adhesion promoter in natural fiber reinforced powder polypropylene composites. *J. Mater. Sci. Lett.* **2002**, *21*, 1885–1888. [CrossRef]
102. Mwaikambo, L.Y.; Ansell, M.P. Chemical modification of hemp, sisal, jute, and kapok fibers by alkalization. *J. Appl. Polym. Sci.* **2002**, *84*, 2222–2234. [CrossRef]
103. Corrales, F.; Vilaseca, F.; Llop, M.F.; Gironès, J.; Méndez, J.A.; Mutjé, P. Chemical modification of jute fibers for the production of green-composites. *J. Hazard. Mater.* **2007**, *144*, 730–735. [CrossRef]
104. Basak, R.; Choudhury, P.; Pandey, K.M. Impacts of Temperature Disparity on Surface Modification of Short Jute Fiber-Reinforced Epoxy Composites. *IOP Conf. Ser. Mater. Sci. Eng.* **2017**, *225*, 12114. [CrossRef]
105. Thakur, M.K.; Gupta, R.K.; Thakur, V.K. Surface modification of cellulose using silane coupling agent. *Carbohydr. Polym.* **2014**, *111*, 849–855. [CrossRef] [PubMed]
106. Gassan, J.; Bledzki, A.K. Effect of cyclic moisture absorption desorption on the mechanical properties of silanized jute-epoxy composites. *Polym. Compos.* **1999**, *20*, 604–611. [CrossRef]
107. Battegazzore, D.; Frache, A.; Carosio, F. Sustainable and High Performing Biocomposites with Chitosan/Sepiolite Layer-by-Layer Nanoengineered Interphases. *ACS Sustain. Chem. Eng.* **2018**, *6*, 9601–9605. [CrossRef]
108. Vlaev, L.; Turmanova, S.; Dimitrova, A. Kinetics and thermodynamics of water adsorption onto rice husks ash filled polypropene composites during soaking. *J. Polym. Res.* **2009**, *16*, 151–164. [CrossRef]
109. Benard, P.; Kroener, E.; Vontobel, P.; Kaestner, A.; Carminati, A. Water percolation through the root-soil interface. *Adv. Water Resour.* **2016**, *95*, 190–198. [CrossRef]
110. Wang, Y.; Wei, Q.; Wang, S.; Chai, W.; Zhang, Y. Structural and water diffusion of poly(acryl amide)/poly(vinyl alcohol) blend films: Experiment and molecular dynamics simulations. *J. Mol. Graph. Model.* **2017**, *71*, 40–49. [CrossRef]
111. Claus, H. Laccases: Structure, reactions, distribution. *Micron* **2004**, *35*, 93–96. [CrossRef]
112. Liu, R.; Dong, A.; Fan, X.; Yu, Y.; Yuan, J.; Wang, P.; Wang, Q.; Cavaco-Paulo, A. Enzymatic Hydrophobic Modification of Jute Fibers via Grafting to Reinforce Composites. *Appl. Biochem. Biotechnol.* **2016**, *178*, 1612–1629. [CrossRef] [PubMed]
113. Riva, S. Laccases: Blue enzymes for green chemistry. *Trends Biotechnol.* **2006**, *24*, 219–226. [CrossRef] [PubMed]
114. Dong, A.; Yu, Y.; Yuan, J.; Wang, Q.; Fan, X. Hydrophobic modification of jute fiber used for composite reinforcement via laccase-mediated grafting. *Appl. Surf. Sci.* **2014**, *301*, 418–427. [CrossRef]
115. Hu, R.-H.; Sun, M.-Y.; Lim, J.-K. Moisture absorption, tensile strength and microstructure evolution of short jute fiber/polylactide composite in hygrothermal environment. *Mater. Des.* **2010**, *31*, 3167–3173. [CrossRef]
116. Hong, C.K.; Kim, N.; Kang, S.L.; Nah, C.; Lee, Y.-S.; Cho, B.-H.; Ahn, J.-H. Mechanical properties of maleic anhydride treated jute fibre/polypropylene composites. *Plast. Rubber Compos.* **2008**, *37*, 325–330. [CrossRef]
117. Mouritz, A.P.; Gibson, A.G. *Fire Properties of Polymer Composite Materials*; Springer Science & Business Media: Berlin, Germany, 2007.
118. Dasari, A.; Yu, Z.-Z.; Cai, G.-P.; Mai, Y.-W. Recent developments in the fire retardancy of polymeric materials. *Prog. Polym. Sci.* **2013**, *38*, 1357–1387. [CrossRef]
119. Chapple, S.; Anandjiwala, R. Flammability of Natural Fiber-reinforced Composites and Strategies for Fire Retardancy: A Review. *J. Thermoplast. Compos. Mater.* **2010**, *23*, 871–893. [CrossRef]
120. Wang, W.; Sain, M.; Cooper, P. Hygrothermal weathering of rice hull/HDPE composites under extreme climatic conditions. *Polym. Degrad. Stab.* **2005**, *90*, 540–545. [CrossRef]

121. A Review: Natural Fiber Composites Selection in View of Mechanical, Light Weight, and Economic Properties Ahmad 2014 Macromolecular Materials and Engineering—Wiley Online Library. Available online: http://onlinelibrary.wiley.com/doi/10.1002/mame.201400089/full (accessed on 7 December 2017).
122. Hollaway, L. A review of the present and future utilisation of FRP composites in the civil infrastructure with reference to their important in-service properties. *Constr. Build. Mater.* **2010**, *24*, 2419–2445. [CrossRef]
123. Stark, N.M.; White, R.H.; Mueller, S.A.; Osswald, T.A. Evaluation of various fire-retardants for use in wood flour–polyethylene composites. *Polym. Degrad. Stab.* **2010**, *95*, 1903–1910. [CrossRef]
124. Ngo, T.D.; Ton-That, M.T.; Hu, W. Innovative and Sustainable Approaches to Enhance Fire Resistance of Cellulosic Fibers for Green Polymer Composites. *SAMPE J.* **2013**, *49*, 31–37.
125. Manfredi, L.B.; Rodriguez, E.; Wladyka-Przybylak, M.; Vazquez, A. Thermal Properties and Fire Resistance of Jute-Reinforced Composites. *Compos. Interfaces* **2010**, *17*, 663–675. [CrossRef]
126. Horrocks, A.R.; Kandola, B.K.; Kandola, B. Flammability and fire resistance of composites. In *Design and Manufacture of Textile Composites*; Elsevier: Amsterdam, The Netherlands, 2005; pp. 330–363.
127. Sinha, E.; Rout, S.K. Influence of fibre-surface treatment on structural, thermal and mechanical properties of jute fibre and its composite. *Bull. Mater. Sci.* **2009**, *32*, 65–76. [CrossRef]
128. Nam, T.H.; Ogihara, S.; Nakatani, H.; Kobayashi, S.; Song, J.I. Mechanical and thermal properties and water absorption of jute fiber reinforced poly(butylene succinate) biodegradable composites. *Adv. Compos. Mater.* **2012**, *21*, 241–258. [CrossRef]
129. Suppakarn, N.; Jarukumjorn, K. Mechanical properties and flammability of sisal/PP composites: Effect of flame retardant type and content. *Compos. Part B Eng.* **2009**, *40*, 613–618. [CrossRef]
130. Sain, M.; Park, S.; Suhara, F.; Law, S. Flame retardant and mechanical properties of natural fibre–PP composites containing magnesium hydroxide. *Polym. Degrad. Stab.* **2004**, *83*, 363–367. [CrossRef]
131. Morgan, A.B.; Gilman, J.W. An overview of flame retardancy of polymeric materials: Application, technology, and future directions. *Fire Mater.* **2013**, *37*, 259–279. [CrossRef]
132. Fatima, S.; Mohanty, A.; Mohanty, A. Acoustical and fire-retardant properties of jute composite materials. *Appl. Acoust.* **2011**, *72*, 108–114. [CrossRef]
133. Battegazzore, D.; Alongi, J.; Duraccio, D.; Frache, A. Reuse and Valorisation of Hemp Fibres and Rice Husk Particles for Fire Resistant Fibreboards and Particleboards. *J. Polym. Environ.* **2018**, *26*, 3731–3744. [CrossRef]
134. Matkó, S.; Toldy, A.; Keszei, S.; Anna, P.; Bertalan, G.; Marosi, G. Flame retardancy of biodegradable polymers and biocomposites. *Polym. Degrad. Stab.* **2005**, *88*, 138–145. [CrossRef]
135. Hapuarachchi, T.D.; Ren, G.; Fan, M.; Hogg, P.J.; Peijs, T. Fire Retardancy of Natural Fibre Reinforced Sheet Moulding Compound. *Appl. Compos. Mater.* **2007**, *14*, 251–264. [CrossRef]
136. Schartel, B.; Hull, T.R.; Hull, R. Development of fire-retarded materials—Interpretation of cone calorimeter data. *Fire Mater.* **2007**, *31*, 327–354. [CrossRef]
137. Bajpai, P.K.; Ahmad, F.; Chaudhary, V.; Martínez, L.M.T.; Kharissova, O.V.; Kharisov, B.I. *Processing and Characterization of Bio-Composites*; Springer: Berlin, Germany, 2017; pp. 1–18.
138. Biocomposites Reinforced with Natural Fibers: 2000–2010—ScienceDirect. Available online: http://www.sciencedirect.com/science/article/pii/S0079670012000391 (accessed on 18 December 2017).
139. Le Duc, A.; Vergnes, B.; Budtova, T. Polypropylene/natural fibres composites: Analysis of fibre dimensions after compounding and observations of fibre rupture by rheo-optics. *Compos. Part Appl. Sci. Manuf.* **2011**, *42*, 1727–1737. [CrossRef]
140. Bledzki, A.; Jaszkiewicz, A.; Murr, M.; Sperber, V.; Lützendgrf, R.; Reußmann, T.; Lützendorf, R. Processing techniques for natural- and wood-fibre composites. In *Properties and Performance of Natural-Fibre Composites*; Elsevier: Amsterdam, The Netherlands, 2008; pp. 163–192.
141. Bledzki, A.K.; Specht, K.; Cescutti, G.; Müssig, M. Comparison of different compounding processes by an analysis of fibres degradation. In Proceedings of the 3rd International Conference on Eco-Composites, Stockholm, Sweden, 20–21 June 2005; p. 3.
142. Chaitanya, S.; Singh, I. Processing of PLA/sisal fiber biocomposites using direct-and extrusion-injection molding. *Mater. Manuf. Process.* **2017**, *32*, 468–474. [CrossRef]
143. Huda, M.; Drzal, L.; Ray, D.; Mohanty, A.; Mishra, M. Natural-fiber composites in the automotive sector. In *Properties and Performance of Natural-Fibre Composites*; Elsevier: Amsterdam, The Netherlands, 2008; pp. 221–268.

144. Azaman, M.; Sapuan, S.; Sulaiman, S.; Zainudin, E.; Abdan, K.; Sapuan, M.S. An investigation of the processability of natural fibre reinforced polymer composites on shallow and flat thin-walled parts by injection moulding process. *Mater. Des.* **2013**, *50*, 451–456. [CrossRef]
145. Leong, Y.; Thitithanasarn, S.; Yamada, K.; Hamada, H. Compression and injection molding techniques for natural fiber composites. In *Natural Fibre Composites*; Elsevier: Amsterdam, The Netherlands, 2014; pp. 216–232.
146. Thakur, V.K.; Thakur, M.K. Processing and characterization of natural cellulose fibers/thermoset polymer composites. *Carbohydr. Polym.* **2014**, *109*, 102–117. [CrossRef]
147. Stringer, L. Optimization of the wet lay-up/vacuum bag process for the fabrication of carbon fibre epoxy composites with high fibre fraction and low void content. *Composites* **1989**, *20*, 441–452. [CrossRef]
148. Yuhazri, M.; Sihombing, H. A comparison process between vacuum infusion and hand lay-up method toward kenaf/polyester composite. *Int. J. Basic Appl. Sci.* **2010**, *10*, 63–66.
149. Rouison, D.; Sain, M.; Couturier, M. Resin transfer molding of natural fiber reinforced composites: Cure simulation. *Compos. Sci. Technol.* **2004**, *64*, 629–644. [CrossRef]
150. Kang, M.K.; Lee, W.I.; Hahn, H. Formation of microvoids during resin-transfer molding process. *Compos. Sci. Technol.* **2000**, *60*, 2427–2434. [CrossRef]
151. Peng, X.; Fan, M.; Hartley, J.; Al-Zubaidy, M. Properties of natural fiber composites made by pultrusion process. *J. Compos. Mater.* **2012**, *46*, 237–246. [CrossRef]
152. Akil, H.M.; Santulli, C.; Sarasini, F.; Tirillò, J.; Valente, T. Environmental effects on the mechanical behaviour of pultruded jute/glass fibre-reinforced polyester hybrid composites. *Compos. Sci. Technol.* **2014**, *94*, 62–70. [CrossRef]
153. Almeida, J.H.S., Jr.; Amico, S.C.; Botelho, E.C.; Amado, F.D.R. Hybridization effect on the mechanical properties of curaua/glass fiber composites. *Compos. Part B Eng.* **2013**, *55*, 492–497. [CrossRef]
154. Nunna, S.; Chandra, P.R.; Shrivastava, S.; Jalan, A. A review on mechanical behavior of natural fiber based hybrid composites. *J. Reinf. Plast. Compos.* **2012**, *31*, 759–769. [CrossRef]
155. Boopalan, M.; Niranjanaa, M.; Umapathy, M. Study on the mechanical properties and thermal properties of jute and banana fiber reinforced epoxy hybrid composites. *Compos. Part B Eng.* **2013**, *51*, 54–57. [CrossRef]
156. Fiore, V.; Calabrese, L. Effect of Stacking Sequence and Sodium Bicarbonate Treatment on Quasi-Static and Dynamic Mechanical Properties of Flax/Jute Epoxy-Based Composites. *Materials* **2019**, *12*, 1363. [CrossRef]
157. Jawaid, M.; Khalil, H.A.; Abu Bakar, A.; Khanam, P.N. Chemical resistance, void content and tensile properties of oil palm/jute fibre reinforced polymer hybrid composites. *Mater. Des.* **2011**, *32*, 1014–1019. [CrossRef]
158. Jawaid, M.; Khalil, H.A.; Abu Bakar, A. Mechanical performance of oil palm empty fruit bunches/jute fibres reinforced epoxy hybrid composites. *Mater. Sci. Eng. A* **2010**, *527*, 7944–7949. [CrossRef]
159. Shanmugam, D.; Thiruchitrambalam, M. Static and dynamic mechanical properties of alkali treated unidirectional continuous Palmyra Palm Leaf Stalk Fiber/jute fiber reinforced hybrid polyester composites. *Mater. Des.* **2013**, *50*, 533–542. [CrossRef]
160. Scalici, T.; Badagliacco, D.; Enea, D.; Alaimo, G.; Valenza, A.; Fiore, V. Aging resistance of bio-epoxy jute-basalt hybrid composites as novel multilayer structures for cladding. *Compos. Struct.* **2017**, *160*, 1319–1328.
161. Shubhra, Q.T.; Alam, A.K.M.M.; Beg, M.D.H.; Khan, M.A.; Gafur, M.A. Mechanical and degradation characteristics of natural silk and synthetic phosphate glass fiber reinforced polypropylene composites. *J. Compos. Mater.* **2011**, *45*, 1305–1313. [CrossRef]
162. Yang, Y.; Ota, T.; Morii, T.; Hamada, H. Mechanical property and hydrothermal aging of injection molded jute/polypropylene composites. *J. Mater. Sci.* **2011**, *46*, 2678–2684. [CrossRef]
163. Velmurugan, R.; Manikandan, V. Mechanical properties of palmyra/glass fiber hybrid composites. *Compos. Part A Appl. Sci. Manuf.* **2007**, *38*, 2216–2226. [CrossRef]
164. Ahmed, K.S.; Vijayarangan, S.; Naidu, A. Elastic properties, notched strength and fracture criterion in untreated woven jute–glass fabric reinforced polyester hybrid composites. *Mater. Des.* **2007**, *28*, 2287–2294. [CrossRef]
165. Aquino, E.M.F.; Sarmento, L.P.S.; Oliveira, W.; Silva, R.V. Moisture Effect on Degradation of Jute/Glass Hybrid Composites. *J. Reinf. Plast. Compos.* **2007**, *26*, 219–233. [CrossRef]
166. Selver, E.; Ucar, N.; Gulmez, T. Effect of stacking sequence on tensile, flexural and thermomechanical properties of hybrid flax/glass and jute/glass thermoset composites. *J. Ind. Text.* **2018**, *48*, 494–520. [CrossRef]

167. Manikandan, N.; Morshed, M.N.; Karthik, R.; Al Azad, S.; Deb, H.; Rumi, T.M.; Ahmed, M.R. Improvement of mechanical properties of natural fiber reinforced jute/polyester epoxy composite through meticulous alkali treatment. *Am. J. Curr. Org. Chem.* **2017**, *3*, 9–18.
168. Ku, H.; Wang, H.; Pattarachaiyakoop, N.; Trada, M. A review on the tensile properties of natural fiber reinforced polymer composites. *Compos. Part B Eng.* **2011**, *42*, 856–873. [CrossRef]
169. Shalwan, A.; Yousif, B.; Yousif, B. In State of Art: Mechanical and tribological behaviour of polymeric composites based on natural fibres. *Mater. Des.* **2013**, *48*, 14–24. [CrossRef]
170. Faruk, O. Cars from Jute and Other Bio-Fibers, 2009. Available online: http://docplayer.net/48269780-Cars-from-jute-and-other-bio-fibers.html. (accessed on 28 December 2018).
171. Karus, M.; Kaup, M.; Lohmeyer, D. Study on markets and prices for natural fibres (Germany and EU). In Proceedings of the 3rd International Symposium Bioresource Hemp, Wolfsburg, Germany, 13–16 September 2000.
172. Netravali, A.N.; Chabba, S. Composites get greener. *Mater. Today* **2003**, *6*, 22–29. [CrossRef]

© 2019 by the authors. Licensee MDPI, Basel, Switzerland. This article is an open access article distributed under the terms and conditions of the Creative Commons Attribution (CC BY) license (http://creativecommons.org/licenses/by/4.0/).

MDPI
St. Alban-Anlage 66
4052 Basel
Switzerland
Tel. +41 61 683 77 34
Fax +41 61 302 89 18
www.mdpi.com

Fibers Editorial Office
E-mail: fibers@mdpi.com
www.mdpi.com/journal/fibers

www.ingramcontent.com/pod-product-compliance
Lightning Source LLC
LaVergne TN
LVHW072000080526
838202LV00064B/6799